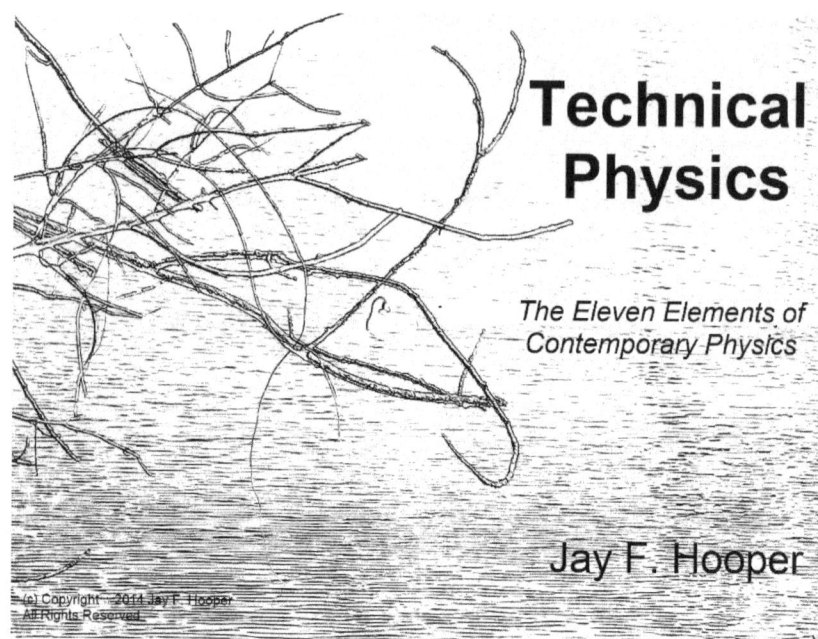

Technical
Physics

*The Eleven Elements of
Contemporary Physics*

Jay F. Hooper

Technical Physics
2nd Edition
The Eleven Elements of Contemporary Physics

Copyright © 2010-2014 by Jay F. Hooper

Printed in the United States of America

ISBN-13: 978-1500342746

ISBN-10: 1500342742

Presented to:

From:

Date:

Table of Contents

1

Forward

This book was written for the university and community college community in order for people to better grasp the underlying concepts and threads in the physical world around us. I have attempted to give a brief, but focused insight into the general realm of Physics and how it relates to our present day contemporary world. Metric values are given within the { } brackets when standard values are used in the text. This is the revised and updated edition that also contains two additional chapters that were not included in the original 2010 book.

Chapter 1: Length and Frequency

When you look at the world of things in dimension 1, in most situations you are greeted with the concepts of length and time. However, you will have a much better understanding of what occurs here if you look at the landscape as one of length and frequency.

Length

In most situations you will be dealing with all standard measurements (feet), or all metric measurements {meters}. In real world situations you would very rarely convert between the two.

In situations where you would need to use measuring devices or to work with different quantities, a handy way to remember equivalencies is through relationships that you experience firsthand. After all, I do not know about you, but I certainly do not always have a measuring tape or ruler with me.

Twelve inches {305 mm} is about equal to the length of one human foot. Two walking paces are about equal to five feet {1.5 meters}. Your outstretched arms are about equal to your height. Four walking paces are about equal to 3 meters (10 feet).

A king size filter cigarette is about 100 mm long {1/10 of a meter}. An inch is about 25 mm long.

3

Standard

There are three countries in the word that use standard measurement, the US, Liberia, and Pongo Pongo. It is not an all-around favorite system. It is a comfortable friend that lives in a society that is somewhat different. It is sort of like a raccoon trying to live on a deer preserve. You can't get rid of the pest, but if he doesn't watch out he will get trampled underfoot.

Metric

The French revolution established this unit of length, and catch on it did. In most instances you will be using meters {m} and millimeters {mm} in your work. One mm is equal to 1/1000 of a meter {.001m}.

Frequency

This is just a fancy way to say per second or per hour, or per minute. An electric motor rotates at 1740 RPM, as in revolutions per minute. That is a frequency (f) or an inverse period (1/T) [inverse time 1/t]. One way to think of this situation is that a frequency is something that repeats over and over and over again that is measurable.

Most of the time (no pun intended) you would run into this concept as Hertz (Hz).

$$f = 1/T$$
$$1740 \text{ RPM} = 1/T$$
$$29 \text{ RPS} = 1/T$$
$$29 \text{ per sec.} = 1/T$$
$$29/\text{sec.} = 1/T$$
$$29 \text{ Hz} = 1/T$$

29 Hz = f
f = 29 Hz
or
f = 1/T
f = 1740 RPM
f = 29 RPS
f = 29 per sec.
f = 29 Hz

Inverse Period

This is just a fancy way to say per second or per hour, or per minute. An electric motor rotates at 1740 RPM, as in revolutions per minute. That is an inverse period (1/T) or a frequency (f). One way to think of this situation is that inverse period is something that repeats over and over and over again, that is measurable.

If you visit Europe you have 50 Hz electric mains. That is a period of .02 seconds.

T = 1/f
T = 1/50 Hz
T = 1/ (50 per sec.)
T = .02/ (per sec.)
T = .02 sec.
then
1/T = 1/ (.02 sec.)
1/T = 50/sec.
1/T = 50 per sec.
1/T = 50 Hz
f = 50 Hz

Inverse Time

This is just a fancy way to say per second or per hour, or per minute. Sound familiar? An electric motor rotates at 1740 RPM, as in revolutions per minute. That is an inverse time $(1/t)$ or a frequency. One way to think of this situation is that inverse time is something that repeats over and over and over again that is measurable. The earth rotates around the sun and we then get one year. An inverse time would be a per year. I have one birthday per year. How about you?

$$t = 1/f$$
$$t = 1/ (1 \text{ per year})$$
$$t = 1/\text{per year}$$
$$t = 1 \text{ year}$$
$$1/t = 1/ (1 \text{ year})$$
$$1/t = 1 \text{ per year}$$
$$\text{or}$$
$$f = 1 \text{ per year}$$

The best way to think about the concepts in this chapter is that you are dealing with length and inverse time (per second) in the dimension 1 area. It just happens to be called length and frequency most of the time. Think of it like this:

dimension 1	dimension 1	
L track	1/T track	
(length)	(per second)	
L	1/T	

	omega	
length	frequency	
ft.	f	
foot	hertz	
m.	Hz.	
meter	1/period	
	1/time	
	rotation	

Chapter 2: Area and Speed

In the arena of concepts in dimension 2 the most common tracts would be L and L/T.

Area

If you take a length times a length, you would end up with an area. If you are dealing with carpet covering a floor, you would end up with square yards (sq. yds.) {sq. meters} of material on your order sheet. When working with the purchase of a new home, you are dealing with a cost per square foot (sq. ft.) {sq. meters}. In dealing with pressurization, you are dealing with a pressure per square inches (sq. in.) {sq. m.}.

While researching possible solar home heating enhancements I once was working with estimates of solar heat loss and gain through windows in units of square meters (sq. ft.).

Speed

If you take a length times an inverse time (1/t) or frequency you end up with speed. I average 60 miles per hour (mph) {97 kph} when I drive up to New Jersey to visit my sister. That is also around 88 feet per sec. (fps) for those of us trying to scurry across a road with our pets in tow.

The speed of sound is around 300 meters per second {mps} [670 mph] while at about a million times faster you have the speed of light at around 300,000 kilometers per second {kps} [186,000 miles/sec] or 300,000,000 mps [1 billion fps].

A neat trick to do during a nearby thunderstorm is to count one, one thousand, two, one thousand, etc. between the interval of a lightning strike and the sound of the thunder from that strike. It turns out that for each one second delay in the sound's arrival time you have about a one thousand foot {300 meters} distance to the stroke (about five seconds per mile) {about 3 seconds per kilometer}. Some of these lightning strikes are pretty darn close.

Velocity

On the factory floor and about everywhere else the terms speed and velocity are used interchangeably, but in reality speed has a magnitude, and velocity has a magnitude and direction. Since most of the concepts and terms that you will be working with use velocity, we will use that term in this text from now on.

L squared or L²

As you can see from the following illustration the L track (length track) from dimension 1 is now expanded to include L squared in dimension 2, known as area.

dimension 1		dimension 2	
L track (length)		L track	
L		L^2	
length		area	
			sq. in.
ft.		sq.ft.	
			square inches
foot		square feet	
			sq. yds.
m.		sq.m.	
			square yards
meter		square meters	

L/T

The T to the minus one track or the 1/T track (per second track) from dimension 1 has now been expanded to include the L/T track in dimension 2, known as velocity.

dimension 1		dimension 2
1/T track (per second)		L/T track (velocity)
1/T		L/T
RPM		speed
frequency		velocity
f		mph
hertz		miles per hour
Hz.		fps
1/period		feet per sec.
1/time		kph
rotation (omega)		kilometers per hour

The entire dimension 2 area would look like this:

dimension 2	dimension 2	dimension 2
--L track--	--L/T track--	--1/T track--
(area)	(velocity)	(per second squared)
L^2	L/T	$1/T^2$
area	speed	omega squared
sq.ft.	velocity	angular acceleration
square feet		
sq. m.	mph	
square meters	miles per hour	
sq. in.	fps	
square inches	feet per sec.	
sq. yds.	kph	
square yards	kilometers per hour	

Chapter 3: Volume, Acceleration, Mass, and Charge

In the dimension 3 area we have a continuation of the L and the L/T track and the beginnings of the two other major tracks, M and Q.

Volume

L^3 or an area times a length gives you a volume. In the standard world we are talking about cubic inches (cu. in.), cubic feet (cu. ft.), cubic yards (cu. yds.), or gallons (gal.). In the metric world you will usually use cubic meters {cu. m.}, cubic centimeters {cc}, milliliters {ml}, or liters {l.}.

Approximate stuff:

There are about 7 and 1/2 gallons in a cubic foot. A liter is about 6 per cent larger than a quart. There are about 1000 cu. ft. in a 3 meter cube.

Exact Stuff:

There are 231 cu. in. in a gallon.

There are 4 quarts in a gallon.

There are 1728 cu. in. in a cu. ft.

There are 27 cu. ft. in a cu. yd.

There are 35.4 cu. ft. in a cubic meter.

There are 1000 cubic centimeters {cc} or {1000 ml} in a cubic liter.

There are 1000 liters in a cubic meter.

There is a farmer near Bellingham, Washington who has a unique feature on his farm. It seems that he was constructing an additional farm building and he was multiplying the square feet of the foundation, times the inches of height in order to get cubic feet of concrete needed. They sell concrete by the cubic yard, so he sort of miscalculated the conversion to cu. yds. from his drawings. The conversation that day went something like this:

The truck driver says, "Well, that does it for this third and last truck on your foundation here. Where do you want the rest of the concrete?"

The farmer replies, "Oh, I don't need any more (concrete)."

Truck driver, "Hey, wait a minute you ordered all this stuff. I can't take it back."

Farmer, "What do you mean?"

Truck driver, "Just that. I can't take it back. Where do you want it?"

Farmer, "Shoot!"

Driver, "Say what?"

Farmer, "Damn."

Driver, "Well I don't know about any dam. How about that old farm pond over there?"

To this day that farmer has the best pond bottom of any farm pond in the western Washington state area.

dimension 3	dimension 3	
L track	L track	
(volume)	continued	
L^3	$(L^2)(L)$	

volume	(area)(length)	
cu.ft.	gal.	
cubic feet	gallons	
cu.m.	cc	
cubic meters	cubic centimeters	
cu.in.	l.	
cubic inches	liter	
cu.yds.		
cubic yards		

Acceleration

When multiplying a velocity times a (per sec.) or an inverse time, you end up with an acceleration. One way of thinking about this is that acceleration is the rate of change of the velocity. Going from 0 to 60 mph {97 kph} in a car in 8 seconds feels a whole lot different than in going from 0 to 60 mph {97 kph} in 20 seconds, as in the neighborhood putt putt mobile.

Gravity

In jumping off any height you just can't seem to escape it. A constant acceleration of 32 feet per second squared (ft./sec. squared) or 9.8 meters per second squared {m./sec.^{2}}. This is when things can go bump in the night or the

14

daytime for that matter. Things can go from bad to worse in a very short time.

When you first jump your downward speed is 0 fps and your acceleration is (32 ft./sec. squared [fps^2]) {9.8 m./sec.^{2}}. At the one second mark your acceleration remains at 32 ft./sec.^{2} {9.8 m./sec.^{2}}, your velocity is now 32 fps {9.8 mps}, and you have fallen 16 ft. {4.9 m}. Ouch! At the two second mark your acceleration remains constant at 32 ft./sec.^{2} {9.8 m./sec.^{2}}, your velocity is 64 fps (about 44 mph) {19.5 mps} and you have fallen 64 ft. {19.5m.}. I do not even want to think about it.

Standard table

acceleration time	velocity result	distance
32 fps^2 0 sec.	0 fps oops	-0 ft.
32 fps^2 .5 sec.	16 fps ouch	-4 ft.
32 fps^2 1 sec.	32 fps damn	-16 ft.
32 fps^2 1.5 sec.	48 fps crunch	-36 ft.

32 fps^2	64 fps	-64 ft.
2 sec.	oh my god	
32 fps^2	96 fps	-144 ft.
3 sec.	splat	

Metric table

acceleration time	velocity result	distance
9.8 mps^2	0 mps	-0 m.
0 sec.	oops	
9.8 mps^2	4.9 mps	-1.2 m.
.5 sec.	ouch	
9.8 mps^2	9.8 mps	-4.9 m.
1 sec.	damn	
9.8 mps^2	14.6 mps	-11 m.
1.5 sec.	crunch	
9.8 mps^2	19.5 mps	-19.5 m.
2 sec.	oh my god	
9.8 mps^2	29.3 mps	-43.9 m.
3 sec.	splat	

dimension 3	dimension 3		dimension 3
L track	L track		L/T track
(volume)	continued		(acceleration)
L^3	$(L^2)(L)$		L/T^2
length cubed	(area) times (length)		
cu. ft.	gal.		a
cubic feet	gallons		acceleration
cu. m.	cc		g
cubic meters	cubic centimeters		gravity
cu.in.	l.		ft./sec.^{2}
cubic inches	liter		feet per second^2
cu. yds.			m./sec.^{2}
cubic yards			meters per second^2

Mass

Most folks know the unit of mass as kilograms {kg.}, but get the concept of mass all mixed up with weight due to the shipping industry and weights and measures in particular. If something is shipped the documentation may list the "weight" in both pounds (lbs.) and kilograms {kg.}. Everybody and his Uncle knows that 2.2 lbs. is equal to 1 kg., but it is not. Not now and never has been. A kilogram {kg.} is a mass and a pound (lb.) is a weight.

So what exactly is a mass? It is a property of a material that exists in the mechanical sense of matter. In most cases the mass of an object for human beings shows up when dealing with gravity and forces. This will be covered later.

Charge

So what exactly is a charge? It is a property of a material that exists in the electrical sense of matter that in most cases for human being shows up when dealing with electricity. We have not covered it yet. The main thing that you need to realize is that all of these concepts can be thought of as a manifestation of the same underlying principle.

dimension 3	dimension 3	dimension 3	dimension 3
L track	L/T track	M track	Q track

volume	acceleration	mass	charge
	gravity		
cu.ft.	ft./sec	kg.	coul.
cubic feet	feet per second	kilograms	coulombs

The dimension 3 area would look something like this:

dimension 3	dimension 3	dimension 3	dimension 3
L track (volume)	L/T track (acceleration)	M track (mass)	Q track (charge)
L^3	L/T^2	M	Q
length cubed	velocity/time		
cu. ft.	a	kg.	coul.
cubic feet	acceleration	kilograms	coulombs
cu. m.	g		

cubic meters	gravity		
cu.in.	ft./sec.		
cubic inches	feet per second		
gal.	m./sec.		
gallons	meters per second		

Chapter 4: Flow

Within the dimension 4 area the big news is what happens in the L track, the M track, and the Q track when you apply a rate of change (a per second) to the dimension 3 concepts covered above.

Mass Flow

In a large commercial or industrial building when you have an HVAC unit for indoor climate control, one of the things that you are going to do is move a bunch of air around. In a technical sense you are moving x amount of air mass per unit time when the ventilating fan(s) are on. This would be in kilograms per second or kg./sec.

On large commercial airline flights fuel is also burned in mass per unit time, such as kilograms per hour {kg. /hr.} or kilograms per minute {kg./min.}. These would be considered mass flows.

Volume Flow

Another way to look at the same situation as the above example would be as follows:

In a large commercial or industrial building when you have an HVAC unit for indoor climate control, one of the things that you are going to do is move a bunch of air around. In a technical sense you are moving x amount of air volume per unit time when the ventilating fan(s) are on. This would be in standard cubic feet per minute (SCFM) {LPM} or standard cubic feet per second (SCFS) {LPS}.

On large commercial airline flights fuel is also burned in volume per unit time, such as gallons per hour (gal./hr.) {liters/hr.} or gallons per minute (gal./min.) {liters/min.}. These would be considered a volume flow.

In most commercial and industrial work such as plumbing and hydraulics the volume flow is most commonly known as GPM (gallons per minute) {LPM}.

GPM

This is the most common designation of volume flow in industry.

dimension 4		dimension 4	
L track		M track	
L^3/T		M/T	
volume flow		mass flow	
SCFH		kg./hr.	
SCFM		kg./min.	
SCFS		kg./sec.	
GPM			
gallons per minute			
l./sec.			
liters/sec.			

Charge Flow

In the electrical realm a unit of charge Q is known as a coulomb (C). When you have a charge flow (coulombs/sec.) you have what is called amperage, or more specifically one ampere (1 amp) is equal to one coulomb of charge moving from point A to point B in one second. An electron, considered to be the "atomic" unit of charge is equal to about 1.6 (10^{-19}) coulombs.

Amp

This is the most common designation of charge flow in industry.

dimension 4		dimension 4	dimension 4
L track		M track	Q track
L³/T		M/T	Q/T
volume flow		mass flow	charge flow
SCFH		kg./hr.	coul./sec
SCFM		kg./min.	coulomb per second
SCFS		kg./sec.	ampere
	l./sec.		
GPM			amp
	liters per second		
gallons per minute			

Tesseract

If you take a volume (L) (L) (L) or L³ commonly known as a cube and you multiply it by a length (L), you will end up with a four dimensional cube called a Tesseract. This object can be thought of in a lot of different ways but one of the industrial ways to think of it, is as an extrusion of sorts. You end up with a Tesseract that is defined or bounded by a three dimensional volume eight times the volume of your original cube.

All of the vertices, lines, sides, etc. of this object can easily be described using regular math formulas.

Chapter 5: Momentum

In the arena of dimension 5, we have the concept of momentum. Generally speaking unless you are reconstructing things like traffic accidents or designing bullet proof vests, you are not going to run into many applications for momentum, except if you shingle a roof. So, what is momentum? Well, it is a mass times a velocity.

If you are playing football and you have a player that has a mass of 100 kg. and he is walking at 2 meter/sec. [4.5 mph], then he would have a momentum of 200 kg.m./sec. If on the other hand that same football player was running at you at 17 meters/sec. [38 mph], then he would have a momentum of 1700 kg.m./sec.

What is the difference between momentums of 200 kg.m./sec. and 1700 kg.m./sec.? Well it is the difference between bumping abruptly into someone and saying, "Oops or excuse me." at 200 kg.m./sec. or being bowled over hard and possibly getting several broken bones if you are unprotected at 1700 kg.m./sec.. A similar thing comes into play with bullets and bullet proof vests. A bullet has a high velocity on the order of say 1000 mps (3300 fps) [2230 mph], but a very low mass on the order of grams (fractional ounces). The football player mentioned previously had a high mass but a low velocity (speed).

In many situations the momentum of that football player walking into you is not that much different than the momentum of a bullet when it strikes you. When a person bumps into you the point of contact is spread over a large surface area, but unfortunately that is not the case with a bullet. The designer of a properly working bullet proof vest must find some way to "spread out" that bullet contact area or lower the velocity in order to save the police officer in a shootout.

If you are putting shingles on a roof, the concepts of momentum also come into play. If you compare the mass of a hammer to that of a typical roofing nail, there is a world of difference. A person who is used to roofing for a living can put a nail in with just two hits or even one for that matter. A novice on the other hand is going to go bam, bam, bam, bam, bam. Even though there is a large mass difference between a hammer and a nail, the higher velocity of the experienced user's swing gives rise to a large momentum when hitting nails.

The multi-dimension chart through dimension 5 would look like this:

dimen-sion 1	dimen-sion 2	dimen-sion 3	dimen-sion 4	dimen-sion 5
length		volume		momen-tum
	speed		kg./sec.	
freq.		gallon		
	velocity		SCFM	
inch		cu. ft.		
	sq. in.		GPM	
foot		cu. yd.		
	sq. ft.		amp	
meter		acceler-ation		

	mph		flow	
L		gravity		
	ft./sec.		M/T	
1/T		mass		
	L/T		Q/T	
		charge		
	area			
		M		
	square meter			
		Q		
	kph			
	(L)(L)			

Chapter 6: Force and Weight

Within the dimension 6 area we come upon two very important concepts, force and weight. They are united by something that is very familiar to most people, pounds (lbs.). It so happens in the world of Physics that:

$$F = ma$$

(force is equal to mass times acceleration) and that: $$W = mg$$

(weight is equal to mass times the acceleration of gravity)

The units of force (F) and weight (W) are lbs. [or newtons].

It is also true that gravity (g) is acceleration and that:

$$g = 32 \text{ ft./sec}^2$$

(32 feet per second squared)

or $g = 9.8 \text{ m./sec}^2$

{9.8 meters per second squared}

Weight

In a gravitational situation (on a planet for example), any object having a mass would be subject to a gravitational acceleration (g) which would manifest itself as the thing that we perceive as weight. The bowling bowl in the movie "The Christmas Story" has a lot of weight but the wrapping paper that it came in has hardly any weight at all. The reason that this is true is that even though everything is experiencing the same acceleration (g) the bowling bowl has a lot more mass then the wrapping paper.

Force

If you accelerate an object using a car, a rocket, or whatever, it experiences an acceleration that we humans experience as an added weight (over and above the 170 lbs. {756 newtons} that we already feel as the average person from our planetary perceptions). The most wild and wooly acceleration that must people experience is a roller coaster ride, but some car rides can be plenty exciting too.

Pounds

Most people experience weight and force through the perception of pounds (lbs.), while in the metric system there are newtons {N or Nt.} to consider. The average person can tell you what a couple of pounds of weight of this or that are like, but sometimes they do not realize that lbs. are a manifestation of either, force or weight, or both. The multi-dimension, multi-relational chart through dimension 6 would look like this:

dimension 1	dimension 2	dimension 3	dimension 4	dimension 5	dimension 6
length		volume		momentum	
	speed		kg./sec.		weight
freq.		gallon			
	velocity		SCFM		force
inch		cu. ft.			
	sq.in.		GPM		pounds
foot		cu. yd.			
	sq. ft.		amp		lbs.
meter		acceleration			
	mph		flow		newton

L		gravity			
	ft./sec.		M/T		
1/T		mass			
	L/T		Q/T		
		charge			
	area				
		M			
	square meter				
		Q			
	kph				
	L^2				

Chapter 7: Energy & Work

Within the dimension 7 area we come upon the physical term "work", but by and large you can think of this area is one of energy.

Work

In applying a force through a certain distance you have done work or:

$$W = F \, s$$

(work equals force times distance)

So we are talking about ft.lbs. (foot pounds) or oz.in. (ounce inches) or in metric, Nm {newton meters}. When pushing against that hypothetical airplane passenger that weighs 170 lbs. {756 Nt.}, if we are able to push them 10 ft. {about 3 m.} down the hall then we have done 1700 ft.lbs. {2300 Nm} of work (170 lbs. times 10 ft.) {756 Nt. times 3.05 m.}.

As one minister so apply put it during a service, "Work for the night is coming when man's work is through". Well I do not know about that, but those units look mighty familiar.

Torque

In tightening nuts unto bolts, or when you have an electric motor you have motion around a rotating point which results in ft.lbs, oz.in, or Nm. This is called torque. If you tighten a nut with a one foot long wrench and you weigh 170 lbs. {756 Nt.} then the nut experiences 170 ft.lbs. {230 Nm} of torque. Torque has the same units as work.

Calorie

If you heat up a pail of water or raise the temperature of a body you must expend so many calories. An average person may consume 2000 calories of food a day. Well, I wonder how much that jelly donut contains?

BTU

By the time we heat up rooms in the HVAC area we start talking about how many BTUs (British thermal units) it takes. So how many

BTUs does that old room air conditioner that is hanging out your window have?

Tons

When the HVAC folks start cooling buildings they start talking about how many tons of cooling capacity that they need. The actual conversions for all of these will be in the Appendix in the back of the book.

Energy

All of the before mentioned items are a manifestation of the concept of energy. The unit of energy is joules. The most difficult part for most people to understand about energy is that it is blind to time. It is only sensitive to the task at hand. An example: A three story building has both a stair and an elevator. A group of six people goes to the top of the building on three different occasions.

(1) They ride the elevator up and it takes 10 seconds.

(2) They walk up the stairs and it takes 3 minutes.

(3) They run up the stairs and it takes 1 minute.

Each time it took exactly the same amount of energy to get the six people to the top of the three story building.

What that means is that whether I ride my bike around the lake and the neighborhood in the housing development where I live (a distance of 6 1/2 miles {10.5 km.} or 45 minutes) or whether I walk it, or whether I jog it, I burn the same amount of calories each time. No matter how I complete the circuit in whatever time frame, doing whatever type of physical activity (bike, hike, jog, or run); I burn exactly the same amount of calories.

So does walking or riding a bike help muscle tone? You bet. Energy is blind to time; it is only sensitive to the task at hand. The multi-dimension, multi-relational chart through dimension 7 would look like this:

dimension 1	dimension 2	dimension 3	dimension 4	dimension 5	dimension 6	dimension 7
length		volume		momentum		work
	speed		kg./sec.		weight	
freq.		gallon				torque
	velocity		SCFM		force	
inch		cu. ft.				calorie
	sq.in.		GPM		lbs.	
foot		cu. yd.				BTU
	sq. ft.		amp		lbs.	

D1	D2	D3	D4	D5	D6	D7
meter		acceler-ation				energy
	mph		flow		new-ton	
L		gravity				joule
	ft./sec.		M/T			
1/T		mass				
	L/T		Q/T			
		charge				
	area					
		M				
	square meter					
	L²					
		Q				
	kph					
D1	D2	D3	D4	D5	D6	D7

Chapter 8: Power

Within the dimension 8 area you can think of this arena as one of the rate of change of energy or more simply put, power.

Watt

This is the unit of power, which is a joule per second {j/sec.}. Most people know this terminology from electric light bulbs or electric heaters. For example, suppose you have two 150 watt bulbs and hooked them up to a generator, which in turn is hooked to a bicycle. It would be tough for one person to light these bulbs fully bright, pumping away at the pedals because the average person only puts out a maximum of around 1/3 horsepower (250 watts) under stress.

Horsepower

This is another unit of power which is 746 joules per second {j/sec.} or 746 watts. Most people know this terminology from electric motors (H.P. or kW). The two handiest conversions for horsepower are: the electric one, 1 H.P. = 746 watts and the mechanical one, 1 H.P. = 550 ft.lbs./sec. {745 Nm/sec.}

What do these mean and how do they relate? Well if you have ten 75 watt electric light bulbs it would take a 1 H.P. {.75 kW} generator connected to some sort of power source to fully light them up. On the other hand, if you had a one cubic foot block of steel (which weighs about 550 lbs. {about 2500 Nt.}) and raised it one foot {about .3 meter} in the air in one second, it would take a 1 H.P. {.75 kW} motor to do that.

The beauty of knowing about the mechanical conversion of horsepower with feet, pounds, and seconds is that it allows you to trade off one element for the other. For instance if you have the same 1 H.P. {.75 kW} motor used in the last illustration and wanted to move a 55 lb. {245 Nt.} block of metal, you could raise it ten feet {3.05 meters} in the air in one second. Or you could move a 2200 lb. {9790 Nt.} block of metal three inches (1/4 foot) {76 mm} in the air in one second.

In all three of these instances you have 550 ft.lbs. /sec. (1 H.P. or 746 watts) {.746 kW}.

1) (550 lbs.)(1 foot) in 1 second = 550 ft.lbs. /sec. = 1 H.P.
{2500 Nt. x .3m. x 1 sec. = .75 kW}

2) (55 lbs.)(10 feet) in 1 second = 550 ft.lbs. /sec. = 1 H.P.
{250 Nt. x 3m. x 1 sec. = .75 kW}

3) (2200 lbs.)(1/4 foot) in 1 second = 550 ft.lbs. /sec. = 1 H.P.
{9790 Nt. x .076 m. = .75 kW}

The multi-dimension, multi-relational chart through dimension 8 would look like this:

dim 1	dim 2	dim 3	dim 4	dim 5	dim 6	dim 7	dim 8
length		vol-ume		mo-men-tum		work	
	speed		kg./sec.		weight		power
freq.		gallon				torque	
	veloc-ity		SCFM		force		watt
inch		cu. ft.				calorie	
	sq.in.		GPM		lbs.		horse
foot		cu. yd.				BTU	power
	sq. ft.		amp		lbs.		
meter		accel-eration				energy	
	mph		flow		new-ton		
L		gravity				joule	
	ft./sec.		M/T				
1/T		mass					

D1	D2	D3	D4	D5	D6	D7	D8
	L/T		Q/T				
		charge					
	area						
		M					
	sq. meter						
	L²						
		Q					
	kph						
D1	D2	D3	D4	D5	D6	D7	D8

Chapter 9: Resistance

Resistance is in dimension zero.

dim. 0	d l m. 1	d l m. 2	d l m. 3	d l m. 4	d l m. 5	d l m. 6	d l m. 7	d l m. 8
resis-tance								
ohms								

DC Resistance (R)

Resistance R has the units of ohms Ω (omega). The resistance to flow is defined in DC electricity and in purely AC resistive heating as Ohms law, E= IR (voltage = current x resistance).

AC resistance (Z)

Impedance Z has the units of ohms Ω (omega). The resistance to flow is defined in AC electricity as the impedance law, E=IZ (voltage = current x impedance).

Inductive Reactance (X_L)

Inductive reactance X_L has the units of ohms Ω (omega). The restriction of AC current flow through a coil of wire is defined in electricity in the impedance law as inductive reactance. Another way of looking at it is: The restriction of current flow into and out of an inductor is defined in electricity in the impedance law as inductive reactance and has the units of ohms.

$X_L=2\pi f L$, where π = 3.1416 and f = 60 Hz. in this example

Capacitive Reactance (X_C)

The restriction of current flow in a capacitor (charge building up and down on the plates of a capacitor in an AC circuit) is defined in electricity in the impedance law as capacitive reactance and has the units of ohms.

$X_C= 1/(2\pi f L)$, where π = 3.1416 and f = 60 Hz. in this example

Chapter 10: Inductance and Capacitance

This area generally uses four dimensional concepts (such as voltage and current) through eight dimensional concepts and terms (such as power), zero dimensional terms (such as resistance) and minus one dimensional concepts (such as inductance and capacitance) in order to better understand the what and the how of the workings of inductors and capacitors. We will be looking at inductors first.

Voltage (V)

This is defined as electrical pressure. The units of electrical pressure is volts.

Current (I)

This is defined as a flow of charge. The units of electrical flow are amps.

DC resistance (R) and AC resistance (Z) [impedance]

Resistance R and impedance Z has the units of ohms Ω (omega). The resistance to flow is defined in DC electricity and in purely AC resistive heating as Ohms law, E= IR (voltage = current x resistance). The resistance to flow is defined in AC electricity as the impedance law, E= IZ (voltage = current x impedance).

Inductance (L)

The resistance to any change in the current flow through a coil of wire is defined in electricity in Faraday's law and has the units of henrys.

Inductive Reactance (X_L)

The restriction of current flow through a coil of wire is defined in electricity in the impedance law as inductive reactance and has the units of ohms.

$$X_L = 2\pi\,f\,L, \text{ where } \pi = 3.1416$$
and f = 60 Hz. in this example

Power

When determining the amount of energy used in a given circuit in a certain amount of time, you would use power. The units of power are watts.

Series Inductance

Let's look at a typical circuit situation with inductors hooked in series. If you had three inductors like this,

2 H   8 H   20 H

L1   L2   L3

illustration 10-1

hooked up to a 24 volt AC circuit what would the induction in henrys be? When inductors are hooked in series you would add their values together so, 2H + 8H + 20H = 30H or L_t = 30 henrys.

Let's look at a more advanced circuit situation where inductors are hooked in series with just an ohm meter available to read their coil resistance values. If you had three inductors like this with a total resistance value of 4 ohms,

.1 OHMS   1.0 OHMS   2.9 OHMS

$$\underline{\quad \varrho\varrho\varrho \quad \varrho\varrho\varrho \quad \varrho\varrho\varrho \quad}$$

L1   L2   L3

illustration 10-2

hooked up to a 240 volt AC circuit what would the induction in henrys be? Z is 40 ohms for this particular circuit. You have three inductor values L_1, L_2 and L_3. The total circuit impedance Z_t in ohms is equal to the value of the impedance for inductors $Z_t = [(R^2) + (X_L^2)]^{.5}$.

$R = 4$ ohms (This would be the value that an ohm meter would read in illus. 10-2)

so, $R^2 = (4^2) = 16$

and $X_L = 2\pi f L$
$= 2 (3.1416) (60) L$
$= (377 L)$ ohms

then $X_L^2 = (142130 L^2)$

Subbing into $Z = [(R^2) + (X_L^2)]^{.5}$ we have:

$Z^2 = R^2 + X_L^2$

$40^2 = 16 + 142130 L^2$

$142130 L^2 = 1600 - 16$

142130 L^2 = 1584

taking the square root of both sides

377 L = 39.8

L = 39.8/377 = .1055 henrys = L_t
          or
L_1 = approx. .002637 henry
L_2 = approx. .026375 henry
L_3 = approx. .076473 henry

Let's now solve for the total circuit current (I_t). A lot of beginning level people erroneously use Ohms law to get the following answer:

$V_t = (I_t)(R_t)$

substituting

240 = (I_t)(.10 + 1.00 + 2.90)

240 = (I_t)(4)

240/4 = I_t

I_t = 240/4

I_t = 60 amps.

Which is WRONG (Your clamp on ammeter will measure about 6 amps in this circuit not 60 amps.) This is an AC circuit not a DC circuit so you need to use the impedance law E = IZ not Ohms law E = IR. So:

$E_t = (I_t)(Z_t)$

substituting

$240 = (I_t)(40)$

$240/40 = I_t$

$I_t = 240/40$

$I_t = 6$ amps.

Once you have the total circuit current (I_t) you can solve for the voltage drop at each inductor using the impedance law, as the same amount of current will be flowing through each inductor as they are hooked in series.

$Z_t = 40$ ohms, so

$Z1 = 1$ ohm
$Z2 = 10$ ohms
$Z3 = 29$ ohms

$V1 = (I_t)(Z1)$ $V2 = (I_t)(Z2)$

V3 = (I_t)(Z3)

substituting

V1 = (6)(1)   V2 = (6)(10)

V3 = (6)(29)

we have,

V1 = 6 volts   V2 = 60 volts

V3 = 174 volts

To double check your answer just add up the three voltage drops and they should equal the voltage rise of the AC power source (240 volts in this example).

Let's look at another very similar circuit situation with inductors hooked together in a three horsepower (3 HP), three phase motor with the same values we have been using (i.e. Z=40 ohms and R=4 ohms). You have just an ohm meter available to use to read the motor coil resistance values. If you have an inductor coil in a three phase motor with a total resistance value of 4 ohms hooked up to a 240 volt AC circuit what would the induction in henrys be? Z is going to be 40 ohms for this particular circuit. You get that Z value by reading information off of the label of the three phase motor:

Volts = 240

Amps = 6   Then you calculate the impedance in ohms using E=IZ. The answer is 40

Ω (240/6). That means that Z=40 since we have an AC circuit at the three phase motor input.

You have an inductor value L. The total circuit impedance Z_t in ohms is equal to the value of the impedance for inductors, which is $Z = [(R^2) + (X_L^2)]^{.5}$.

R = 4 ohms. (This would be the value that an ohm meter would read at the inputs to the three phase motor coils [L1L2, L1L3, or L2L3]).

so, $R^2 = (4^2) = 16$

and $X_L = 2\pi f L$
$= 2 (3.1416) (60) L$
$= (377 L)$ ohms

then $X_L^2 = (142130 L^2)$

Subbing into $Z = [(R^2) + (X_L^2)]^{.5}$ we have:

$Z^2 = R^2 + X_L^2$

$40^2 = 16 + 142130 L^2$

$142130 L^2 = 40^2 - 16$

$142130 L^2 = 1600 - 16$

$142130 L^2 = 1584$

taking the square root of both sides

377 L = 39.8

L = .1055 henrys

Solve for the total circuit current (I_t). (A lot of people incorrectly use Ohms law E=IR to get the following answer):

$E_t = (I_t)(R_t)$

substituting

$240 = (I_t)(4)$

$240/4 = I_t$

$I_t = 240/4$

$I_t = 60$ amps.

This is WRONG (Your clamp on ammeter will measure about 6 amps in this motor circuit NOT 60 amps.) A 3 HP motor is not ever going to be drawing 60 amps just humming along running some machine unless it is on the verge of blowing up, catching on fire or some other dire set of circumstances. This is an AC motor circuit not a DC circuit so you need to use the impedance law E = IZ not Ohms law E = IR. So:

$E_t = (I_t)(Z_t)$

substituting

$240 = (I_t)(40)$

$240 = (I_t)(40)$

$240/40 = I_t$

$I_t = 240/40$

$I_t = 6$ amps.

To double check your answer just add up the voltage drop and it should equal to the voltage rise of the AC power source (240 volts in this example).

Parallel Inductance

Let's look at a typical circuit situation with inductors hooked in parallel. You have three inductors like this,

illustration 10-3

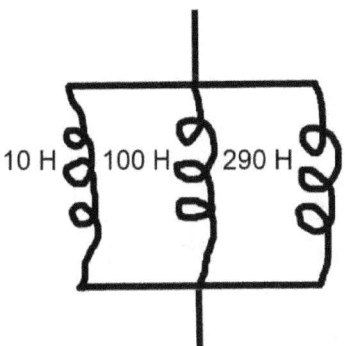

10 H   100 H   290 H

hooked up to a 24 volt AC source. You are given the three inductor values hooked up in parallel

(10, 100, and 290), so you would add them together as reciprocals to get the total circuit induction (L_t). You would add them together like this:

$$1/Lt = 1/L1 + 1/L2 + 1/L3 + \ldots$$

or like this:

$$L_t = \frac{1}{1/L1 + 1/L2 + 1/L3}$$

Using the first equation and solving we have:

$$1/L_t = 1/L1 + 1/L2 + 1/L3$$

$$1/L_t = 1/10 + 1/100 + 1/290$$

$$1/L_t = .1 + .01 + .00345$$

$$1/L_t = .11345$$

$$L_t = 8.815 \text{ henrys}$$

Let's look at a more advanced circuit situation with these same three inductors from illustration 10-3 hooked in parallel, but let the impedance Z be equal to 3.5K ohms for this particular circuit. You have three inductor values L_1, L_2 and L_3. The impedance Z in ohms is equal to the value computed in the total circuit impedance for inductors $Z_t = [(R^2) + (X_L^2)]^{.5}$ which also includes the inductor's resistance value R.

What then would the voltage drop be across each inductor and what would the value of the current going through each inductor be? There are two facts that will help you solve this problem. The sum of the currents flowing through each inductor in the circuit must be equal to the total current flow through the AC power source in the circuit. In other words the voltage drop across each inductor hooked in parallel is the same as the voltage of the AC source.

The voltage drop across each inductor is equal to the voltage rise of the AC source (in this example 24 VAC).

$V_t = V1 = V2 = V3 = 24$ volts

You are given the three inductor values so you would add them together as reciprocals to get the total circuit induction (L_t).

$$L_t = \frac{1}{1/L1 + 1/L2 + 1/L3}$$

$$L_t = \frac{1}{1/10 + 1/100 + 1/290}$$

$$L_t = \frac{1}{.1 + .01 + .00345}$$

$$L_t = \frac{1}{.11345}$$

$L_t = 8.815$ henrys (Look familiar?)

You can double check your answer in the field by utilizing the fact that the total circuit inductance of any number of inductors hooked in parallel will always be less then the value of the smallest inductor in the parallel network. In this case, ask yourself, "Is the computed value of 8.815 henrys for the network less than the 10 henrys of the smallest parallel inductor in this circuit (illus. 10-3)?" The answer is, "Yes it is." So you are OK. You are in the right ballpark.

You know that the total circuit voltage (V_t) is equal to the voltage drop across each inductor in the parallel network. So use the impedance law $E=IZ$ to solve for the total inductor network circuit current flow (I_t).

$$E_t = (I_t)(Z_t)$$

where $Z_t = 3.5K$ ohms of impedance (which is given)

$Z_t = [(R^2) + (X_L^2)]^{.5}$

solving for X_L we have:

$X_L = 2\pi f L$
$= 2 \, (3.1416) \, (60) \, 8.815$
$= (3323) \text{ ohms}$

then $X_L^2 = (11{,}042{,}329) \text{ ohms}^2 = 11 \text{ Meg.}$

subbing into $Z = [(R^2) + (X_L^2)]^{.5}$ we have:

$Z^2 = R^2 + X_L^2$

$(3500)^2 = R^2 + 11 \text{ Meg.}$

$12.25 \text{ Meg.} = R^2 + 11 \text{ Meg.}$

$R^2 = 12.5 \text{ Meg.} - 11 \text{ Meg.}$

$R^2 = 1.25 \text{ Meg.}$

$R = 1118 \; \Omega$

     A unit Z is about equal to 10% more than the network total of 3500 Ω or approx. 3850 Ω. It would be almost exactly equal to [(10+100+290)/3500] larger than 3500, or (400/3500) larger than 3500, or (11.43% larger than 3500) or 3900 Ω. Using the illus. 10-3 relative circuit weights, we have:

Z1 = 1 x 3900 = 3900 Ω

Z2 = 10 x 3900 = 39K Ω

Z3 = 29 x 3900 = 113.1K Ω

substituting into the impedance law
$$E_t = (I_t)(Z_t)$$
we have:
24 = (I$_t$)(3500)

24/3500 = I$_t$

I$_t$ = 24/3500

I$_t$ = .00686 amps = 6.86 ma.

     Once you have the total circuit current (I$_t$) you can solve for the current flow through each inductor using the impedance law E = IZ, as the same amount of voltage will be seen across each inductor, as they are hooked in parallel.

V1 = (I1)(Z1)   V2 = (I2)(Z2)

    V3 = (I3)(Z3)
               substituting
Vt = (I1)(3900)   Vt = (I2)(39K)

Vt = (I3)(113K)

---

24 = (I1)(3900)   24 = (I2)(39K)

24 = (I3)(113.1K)

24/3900 = (I1)   24/39K = (I2)

24/113.1K = (I3)

.0061 = (I1)   .00061 = (I2)

.00021 = (I3)

(I1) = .0061 amps = 6.1 ma.
(I2) = .00061 amps = .61 ma.
(I3) = .00021 amps = .21 ma.

To double check your answer, just add up the three current flows and they should be equal to the current flow of the power supply (6.86 ma. in this example).

Capacitance (C)

The ability of a body to store charge is defined in electricity as capacitance and has the units of farads. Usually you will be using discrete components called capacitors.

Capacitive Reactance (X_C)

The restriction of current flow into and out of a capacitor is defined in electricity in the impedance law as capacitive reactance and has the units of ohms.

$X_C = 1/(2\pi\ f\ L)$, where $\pi = 3.1416$ and f = 60 Hz. in this example of line voltage frequency

Parallel Capacitance

Let's look at a typical circuit situation with capacitors hooked in parallel. If you have three capacitors like this,   illustration 10-4

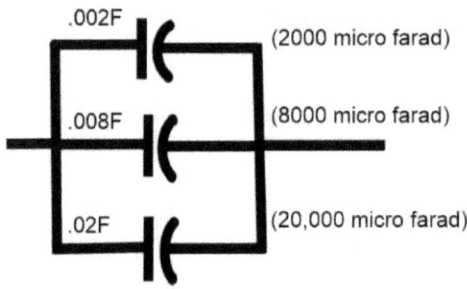

.002F   (2000 micro farad)

.008F   (8000 micro farad)

.02F   (20,000 micro farad)

hooked up to a 24 volt AC circuit what would the capacitance in farads be?

When capacitors are hooked in parallel you would add their values together so that .002F + .008F + .020F = .030 farads or C_t = .030 farads = 30 milli-farads (more commonly called 30,000 µF in the field).

Let's look at another circuit situation with capacitors hooked in parallel. If you have three capacitors like this with a total capacitance value of 4 µF,

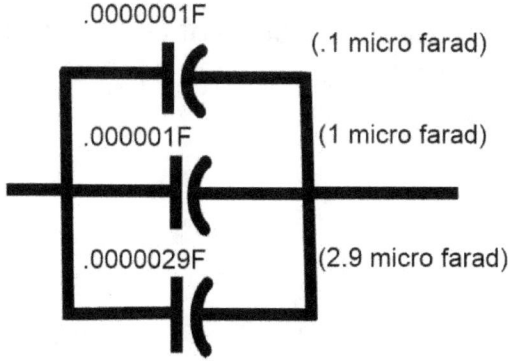

.0000001F

(.1 micro farad)

.000001F

(1 micro farad)

.0000029F

(2.9 micro farad)

illustration 10-5

hooked up to a 24 volt AC circuit, what would the impedance in ohms be? What would the total circuit current be?

Z is unknown for this particular circuit, while the frequency is given as 60 Hz. You have three capacitor values C_1, C_2 and C_3. The total circuit impedance Z_t in ohms = $[(R^2) + (X_L^2 - X_C^2)]^{.5}$ This would be equal to the value of the capacitive reactance X_C since R=0 and X_L = 0 in the circuit in illustration 10-5. There is no coil of wire or any resistor in that circuit so Z = X_C.

X_C = 1/2π f C

= 1/[2 (3.1416) (60) C]

= 1/[(377) (4µF)]

= 1/[(377) (.000004)]

= [1/.001508] ohms

= 663 ohms

Z = 663 ohms
Solving for the total circuit current (I_t) using the
impedance law, we have:
$E_t = (I_t)(Z_t)$

substituting

24 = (I_t)(663)

24/663 = I_t

I_t = 24/663

I_t = .0362 amps or 36.2 ma.

Once you have the total circuit current (I_t)
you know that the voltage drop at each capacitor
is 24 volts. The same amount of voltage will be
dropped across each capacitor as they are
hooked in parallel.
I_t = 36.2 ma., so a unit current I = 36.2/40 = .905
ma.

I_1 = 1 (.905) = .905 ma. = .000905 amps
I_2 = 10 (.905) = 9.05 ma. = .00905 amps
I_3 = 29 (.905) = 26.245 ma. = .026245 amps

To double check your answer just add up
the three flows and they should be equal to the

total current flow of the AC power source (36.2 ma. in this example).

Parallel Capacitance and Inductance

Let's look at a more advanced circuit situation with a capacitor hooked in parallel in a three horsepower (3 HP) single phase capacitor start motor circuit with values R=4 ohms and L =.1055 henry for the coil winding of the single phase motor. If you have a capacitor in a single phase motor circuit with a total capacitive value of 90 µF hooked up in parallel to a 238 volt AC circuit, what would the total circuit impedance Z_t be? Z is going to be about 39.34 ohms for this particular motor. You get that Z value by reading information off the label of the single phase motor:

Volts = 238

Amps = 6.05

Then calculate the impedance ohms using E=IZ. The answer is going to be about 39.34 Ω (238/6.05). That means that impedance Z = 39.34 since we have an AC circuit at the single phase motor.

You also have a 90 µF start capacitor C. The value of the impedance for the run capacitor parallel circuit is $Z = X_C$.

$X_C = 1/2π f C$

$= 1/[2 (3.1416) (60) (.000090)]$

$= 1/.0339$

$= 29.47$ ohms

You have an inductor value L=.1055H The value of the impedance for the motor coil circuit is $Z = [(R^2) + (X_L^2)]^{.5}$ or in this case about 39.34 ohms (from the label).

Double checking we see that:

$X_L = 2\pi f L$
$= 2 (3.1416) (60) .1055$
$= 39.77$ ohms

then $X_L^2 = (39.77)^2 = 1582$
Subbing into $Z = [(R^2) + (X_L^2)]^{.5}$
we have:

$Z^2 = R^2 + X_L^2$
$Z^2 = 16 + 1582 = 1598$
Z = 39.97 ohms (which is very close to the 39.34 ohms that we calculated from the motor label)

Let's look more closely at this motor circuit situation with two branches with impedances hooked in parallel. You have two impedance branches like this,

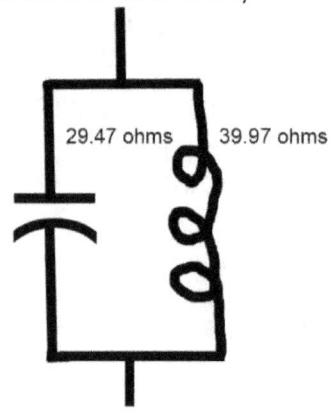

29.47 ohms   39.97 ohms

illustration 10-6

hooked up to a 238 volt AC source. You solved for the two impedance values so you would add them together as reciprocals to get the total circuit impedance (Z_t). You would add them together like this:

$1/Zt = 1/Z1 + 1/Z2 + \ldots$

or like this:

$Z_t = \dfrac{1}{1/Z1 + 1/Z2}$

Using the first of these two equations and solving we have:

$1/Z_t = 1/Z1 + 1/Z2$

Where $Z1 = X_C = 29.47\ \Omega$
and $Z2 = Z = 39.97\ \Omega$

$1/Zt = \qquad 1/Z1 + 1/Z2$

$1/Z_t = \qquad 1/29.47 + 1/39.97$

$1/Z_t = \qquad .03393 + .02050$

$1/Z_t = \qquad .05443$

$Z_t = 18.37\ ohms$

You know that the total circuit voltage (V_t) is equal to the voltage drop across each impedance branch in the parallel network. So use the impedance law $E = IZ$ to solve for the total inductor network circuit current flow (I_t).

$E_t = (I_t)(Z_t)$

where Z_t = 18.37 ohms of impedance

solving for I_t we have:

$238 = (I_t)(18.37)$

$I_t = 238/18.37$

$I_t = 12.956$ amps

The motor will briefly draw this amount of current while starting up until the start capacitor is removed from the circuit. This gives you more torque on startup.

Chapter 11: Permeability and Permittivity

It turns out that permeability and permittivity is in dimension minus 2.

$\frac{d}{dm.}{-2}$	$\frac{d}{dm.}{-1}$	$\frac{d}{dm.}0$	$\frac{d}{dm.}1$	$\frac{d}{dm.}2$	$\frac{d}{dm.}3$	$\frac{d}{dm.}4$	$\frac{d}{dm.}5$	$\frac{d}{dm.}6$	$\frac{d}{dm.}7$	$\frac{d}{dm.}8$
						L^3/T				
ε_o		Ω				M/T				HP
	$\mathbf{L}$		L		M	Q/T	p		P.E	
μ_o		Z		$\mathbf{J}$				F		P
	$\mathbf{C}$		T^{-1}		Q				K.E	

Permittivity ε_o

Permittivity ε_o has the units of farads/meter (Q^2T^2/ML^3). The permittivity of free space is also a constant and is equal to 8.854 10^{-12} farads/meter, [8.854 coul.^{2}/nt.m.^{2}], [8.854 10^{-12} $(Q^2/Ma.L^2)$] or [8.854 10^{-12} (Q^2T^2/ML^3)].

Permeability μ_o

Permeability μ_o has the units of henrys/meter. The permeability of free space μ_o is also a constant and is equal to 4π 10^{-7} henry/meter [4π 10^{-7} (ML/Q^2) or 4π 10^{-7} (volt.T/amp.L) or 12.566 10^{-7} (impedance/velocity).

It turns out that $(\mu_o \, \varepsilon_o)^{-.5}$ is equal to c the speed of light which is another constant in free space. The speed of light $c = (\mu_o \, \varepsilon_o)^{-.5} = (12.566$ $10^{-7}[(8.854 \; 10^{-12}))]^{-.5} = [111.23 \; 10^{-19}]^{-.5} = [11.123$ $10^{-18}]^{-.5} = 1/[3.335 \; 10^{-9}] = 1/3.35 \; 10^{-9} = .2998 \; 10^9 =$ $2.998 \; 10^8$ meters/sec. Where $1/[(Q^2T^2/ML^3)(ML/Q^2)]^{.5} = 1/[T^2/L^2]^{.5} = (L^2/T^2)^{.5} =$ (L/T).

It is also true that $(\mu_o \, c)$ is equal to 377 Ω or Z_o the impedance of free space.
$Z_o = (\mu_o \, c) = [(12.566 \; 10^{-7}) (2.998 \; 10^8)] = [37.67$ $10^1] = [376.7] = 377 \; \Omega$. Where $(ML/Q^2)/(L/T) =$ $ML^2/TQ^2 = (ohm) = \Omega$.

Or put another way:
Z_o or the impedance of free space is equal to 377 Ω or $(\mu_o / \varepsilon_o)^{.5}$. $Z_o = [(12.566 \; 10^{-7}) / (8.854 \; 10^{-12})]^{.5}$ $= [1.41924 \; 10^5]^{.5} = [14.19 \; 10^4]^{.5} = (141900)^{.5} =$ $377 = 377\Omega$

So how did all of this come about and what does it mean for you?

Estimation

For me it was two major epiphanies. The first came about in a college Physics class. You know the type of classroom that you see in the movies, or you may have even experienced it yourself. The blackboards completely surround you on every side in the classroom. My professor, Dr. Antonakos was on the fourth board working his way around the room when he turns and looks at one of the boards in the front of the room and says, "Well, that (meaning that top half of the front board) is about equal to 1/3. Next we will substitute it into this equation here."

I do not know about you, but in all of my math classes through High School it was pretty much drilled into my head that you had to be exact or pretty darn near exact in all of the math equations and problems that you worked on. So, it was a pretty big light bulb going off in my head when I realized that you can estimate a lot of the time and come out pretty close to the "right" answers. I knew about the old saw that says, "Close only counts in horseshoes, hand grenades, and atomic warfare", but this was a big deal to me. Since that time I have used estimation quite frequently in the field in various endeavors as diverse as rigging and fluid mechanics.

Dimensions

The second epiphany came about while walking along some railroad tracks one night. I had donated my car to the local community college a month or two previously, so that they could do repair work on it for practice in their automotive classes. I had been riding my bike to and from work. I had just been hit by a car a week or so previously (but not seriously injured). So, now I was riding the bus to work during the afternoon and walking home from work at night.

As I was walking down the railroad tracks that night, I had come to realize that all of the equations in the engineering texts and in all of the Physics texts that I had encountered were in fact all interrelated to each other. I realized that instead of memorizing all of the equations, or looking up equations for all of the thousands of situations that one could or would run into, you could see them as a relationship instead. You would be able to derive equations and relationships from what you already knew in your head. The first way equations and concepts are related is that they all have M, L, T, and Q units.

dim. -2	dim. -1	dim 0	dim. 1	dim. 2	dim. 3
μ_o					
			L		**L³**
permeability		1		L^2	
	T		**1/T**		L/T^2
ML/Q²		unity		L/T	
	henry				**M**
henry/meter	**L** induc-tance	L to the zero power		v	
	ML²/Q²				
ε_o		T to the zero power			
Q²T²/ML³	farad				
permittivity	**C** capaci-tance				
farad/meter	Q²T²/ML²				

dim. 4	dim. 5	dim. 6	dim. 7	dim. 8
	mv		Fs	
M/T		ma		IV
	p		W	
Q/T		F		P
	p = mv		W = Fs	
		F = ma		P=IV
	ML/T		ML^2/T^2	
		ML/T^2		ML^2/T^3

The second way that they were related was dimensionally. If one assigned the 1st dimension to the L unit, the 1st dimension to the T^{-1} unit and the 3rd dimension to the M and Q unit, then L and $1/T$ would be in the dimension 1 area.

Dim -2	Dim -1	Dim. 0	Dim. 1	Dim. 2	Dim. 3	D4	D5	D6	D7	D8
			L							
		unity								
			1/T							
		1								
			length							
		L to the zero pow-er								
			freq.							
		T to the zero pow-er								
			Hz.							
		resis-tance								
D-2	D-1	D0	D1	D2	D3	D4	D5	D6	D7	D8

L^2 and L/T would be in the dimension 2 area.

Dim -2	Dim -1	Dim. 0	Dim. 1	Dim. 2	Dim. 3	D4	D5	D6	D7	D8
			L							
		unity		L^2						

D-2	D-1	D0	D1	D2	D3	D4	D5	D6	D7	D8
			1/T							
		1		L/T						
			length							
		L to the zero power		v						
			freq.							
		T to the zero power		area						
		resis-tance		speed						
			Hz.							

L^3 and L/T^2 and M and Q would be in the dimension 3 area, while M/T and Q/T would be in the dimension 4 area.

Dim -2	Dim -1	Dim. 0	Dim. 1	Dim. 2	Dim. 3	Dim. 4	Dim 5	Dim 6	Dim 7	Dim 8
			L		L^3					
		1		L^2		M/T				
			1/T		L/T^2					
		unity		L/T		Q/T				
			length		**M**					
		L to the zero power		v		mass flow				
			freq.		**Q**					
		T to the zero power		area		charge flow				
			Hz.		a					
		resis-tance		speed		flow				
					m					
						cur-rent				
					mass					
						volt-age				
					charge					
						PSI				
					accel-eration					
					vol-ume					

Relational Matrix

One could take any term or concept from Physics or engineering, break it down into the resulting M, L, T, and Q units and they would fit on an 11 dimensional matrix from dimension -2 through dimension 8.

Dim -2	Dim -1	Dim. 0	Dim. 1	Dim. 2	Dim. 3
μ_o	T		L	L^2	L^3
		1		area	
perme-ability			1/T		L/T^2
	henry	unity		L/T	a
ML/Q^2			length		
	L induc-tance	L to the zero power		v	**M** mass **Q** charge
henry/ meter			freq.		
	ML^2/Q^2	T to the zero power		speed	
			Hz.		**E**
	C capaci-tance	resis-tance		**J**	
ε_o				Q/L^2T	electric field
	farad		**B**	current	

				density	
Q^2T^2/ML^3		ρ			ML/QT^2
	Q^2T^2/ML^2	rho	mag-netic field		
Permit-tivity		charge density			
			M/QT		
farad/meter		Q/L^3			accel-eration
			tesla		
		Z imped-ance			vol-ume

Dim. 4	Dim. 5	Dim. 6	Dim. 7	Dim. 8
	mv		Fs	
M/T		ma		IxV
	p		W	
Q/T		F		P
	p = mv		W = Fs	
mass flow		F = ma		P=IV
	ML/T		ML^2/T^2	
charge flow		ML/T^2		ML^2/T^3
	momen-tum		work	
flow		force		power
	Φ_E		torque	
current				H.P.

D4	D5	D6	D7	D8
	electric flux		energy	
voltage		$\mathbf{M}$ or $\boldsymbol{\mu}_m$		watts
PSI		mag-netic dipole moment		KW
QL				
$\mathbf{P}$				
electric dipole moment		QL^2/T		
D4	D5	D6	D7	D8

If you had two terms in an equation and they were multiplied together, their dimensional aspects would add (like exponents do).
Take F = ma
Force = (mass) (acceleration)
F = m (mass of dim. 3) times an (acceleration of dim. 3)
F (force) would then be in dim. 6.

If you had two terms in an equation and them divided each other, then their dimensional aspects would subtract (like exponents). Take I = P/V for instance.

current = power/voltage or
flow = power/pressure

I = P (power of dim. 8) divided by V (voltage pressure of dim. 4)

The (current) I would be dim. 4.

The other great thing about this is that it tells you that flow and pressure are of the same dimension. Another way of thinking about it is that they are a manifestation of the same underlying thing.

The end result is an eleven dimensional matrix where all of the terms and concepts of the physical world are shown in their basic relationships with one another.

Dim -2	Dim -1	Dim. 0	Dim. 1	Dim. 2	Dim. 3
μ_o	T		L	L^2	L^3
perme-ability	time	1	length	area	vol-ume
ML/Q^2		unity	$1/T$		L/T^2
	henry		freq.	L/T	a
henry/ meter			Hz.		
	L induc-tance	L to the zero power		v speed	**M** mass **Q** charge
ε_o $Q^2T^2/$ ML^3			**B**		
permit-tivity	$ML^2/$ Q^2	T to the zero power	mag-netic field		$ML/$ QT^2
farad/ meter			M/QT	Q/L^2T	**E**

D-2	D-1	D0	D1	D2	D3
	C capacitance	resistance		**J**	electric field
			current density		
	farad				Φ_B
		ρ			magnetic flux
	Q^2T^2/ML^2	rho			
		charge density			
	σ				
	surface charge	Q/L^3			acceleration
	Q/L^2		tesla		
D-2	D-1	D0	D1	D2	D3

Dim. 4	Dim. 5	Dim. 6	Dim. 7	Dim. 8
	mv		Fs	
M/T		ma		IV
	p		W	
Q/T		F		P
	p = mv		W = Fs	
mass flow		F = ma		P=IV
	ML/T		ML^2/ T^2	
charge flow		ML/T^2		ML^2/ T^3
	momentum		work	
flow		force		power
	Φ_E		torque	
current				H.P.
	electric flux		energy	
voltage		**M** or $\boldsymbol{\mu}_m$		watts
			calorie	
PSI		magnetic dipole moment	BTU joule	KW
P				
electric dipole moment		QL^2/T		
QL				

Chapter 12 Examples In The Mechanical World

This profession is one that is encountered on a daily basis as regards everyday usage (cars, doors, pens, etc.), but it is sometimes a mystery when doing troubleshooting. When working within this area usually the more difficult items to consider are:

Ratio and Proportion

When you need to change the speed of a machine you have a driver pulley, sheave, gear, etc. that is hooked to your prime mover (usually an electric or a gas motor). You also have a driven pulley, sheave, gear, etc. that is hooked to your load. In the case of a car it is your tires.

Ratio and proportion will tell you how fast you are going. So let's look at a bicycle to see how this works.

illustration 12-1

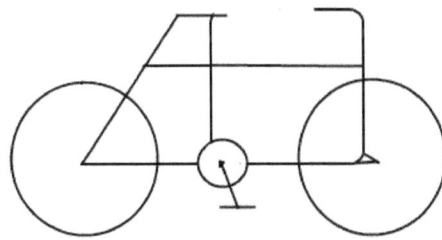

The pedals are hooked to your driver sprocket that is 8 inches {203mm} in diameter. Since this example is a three speed bike our driven sprocket (hooked to the back wheel of the bike) has three choices; a sprocket of 5 inches {127mm} in diameter, a sprocket of 3 inches {76mm} in diameter and a sprocket of 2 inches {51mm} in diameter. In low speed:

illustration 12-2

[Low speed (times 1.6)]

driveN   driveR

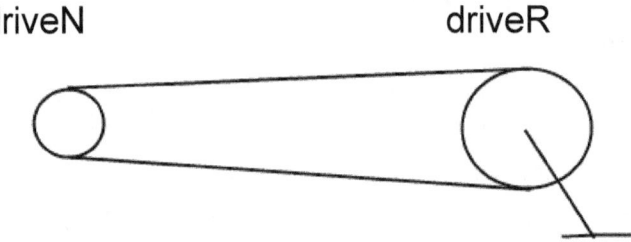

back wheel sprocket   pedal sprocket
5" dia. or {127mm}   8" dia. or {203mm}

The ratio is 8:5 (driver to driven) or to put it another way for every 1 revolution that we push the pedals around 360 degrees our back wheel rotates 1.6 times. In high speed:

illustration 12-3

[High speed (times 4)]
driveN   driveR

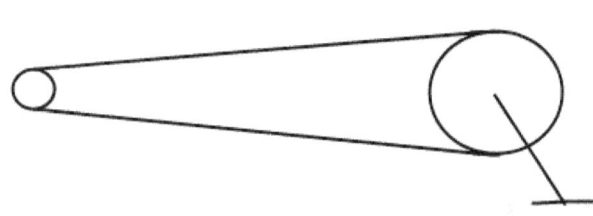

back wheel sprocket   pedal sprocket
2" dia. or {51mm}   8" dia. or {203mm}

The ratio is 8:2 (driver to driven), 4:1, or to put it another way for every 1 revolution that we push the pedals around 360 degrees, our back wheel rotates 4 times. This is the concept of mechanical ratio and proportion. It does not matter in given situation if you compare diameter to diameter.

illustration 12-4

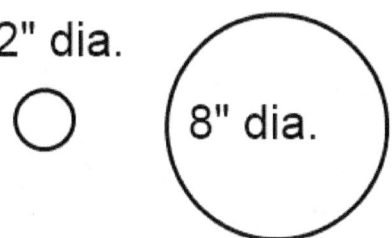

2" dia.

8" dia.

8"/2" {203mm/51mm} = 4 or 4:1

It does not matter in a given situation if you compare radius to radius.

illustration 12-5

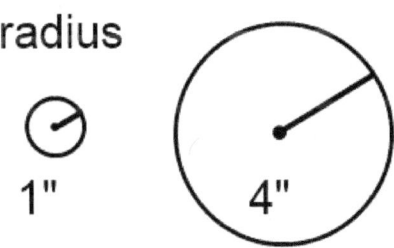

radius

1"

4"

4"/1" {102mm/25.4mm} = 4 or 4:1

It does not matter in a given situation if you compare teeth to teeth.

illustration 12-6

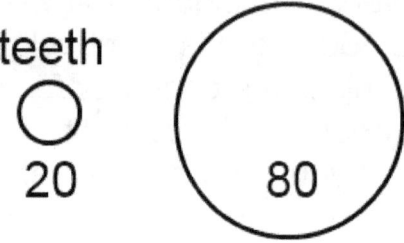

teeth

20   80

80 teeth/20 teeth = 4 or 4:1

Your ratio of driver speed to driven speed will
come out the same.

    With industrial equipment you might typically
have a three phase AC electric motor that turns at
about 1740 RPM. That
motor in turn might be hooked to a speed reducer
with a ratio of 20:1.

                     illustration 12-7

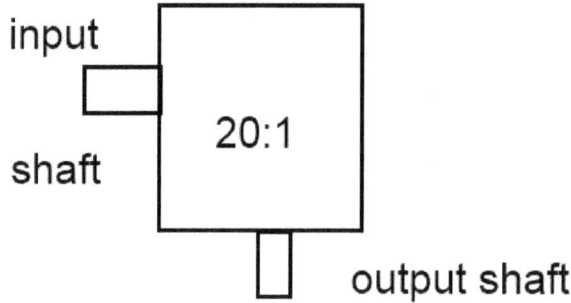

input

shaft

20:1

output shaft

Top view of speed reducer.

    That would mean that the output rotation for
your machine would be around 87 RPM, instead
of a hard to handle 1740 RPM for a conveyor belt
system.

Another big item in the mechanical world is the concept of torque. There are so many nuts and bolts in the world that need to be fastened at the proper torque. Torque is equal to a force in pounds (lbs.) {Nt.} times a distance (usually how long the handle of your wrench is):

Top view of wrench with a 170 lb. {756 Nt.} person about to pull on it. (340 ft.lbs. of torque) {461 Nt.m of torque}

illustration 12-8

2 foot long

wrench

{.61 meters long}

Top view of wrench with a 170 lb. {756 Nt.} person about to pull on it.(1020 ft.lbs. of torque) {1383 Nt.m}

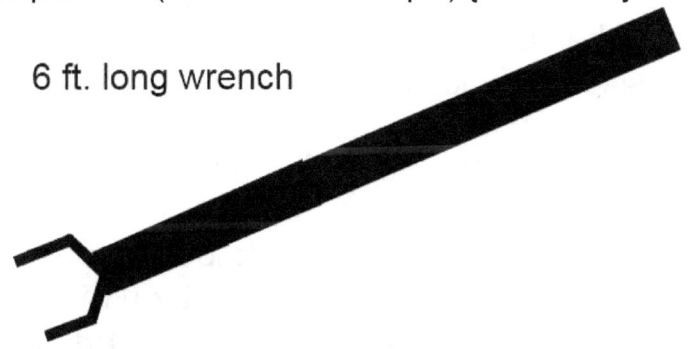

6 ft. long wrench

{1.83 meters long}   illustration 12-9

Oops! A busted bolt. We never even got over 900 ft.lbs. {1220 Nt.m.} of torque in this example as the bolt broke. You guessed it. Cheater bars or cheater pipes are a no no.

The other thing in the mechanical world is how to fix something that is broken. Well, we can take a cue from the world of cars. On average about half of the cars that you see "broken down", at the side of the road only have one of two things wrong with them, spark or gas. What this means is that the engine not receiving either (spark or gas) is at the root cause of half the problems.

You can teach people in a 4 hour seminar how to deal with 50% of all car problems and a triple A card can go a long way for the other ones.

In the mechanical machine world if it is not obvious on visual inspection what the problem is then it can be solved in a systematic way by looking at position, timing, stroke, and other.

Position

A mechanical device needs to be in the proper position to function properly. If it is not, then the arm or linkage of the part of the system involved is either bent or out of whack some-how.

Timing

A mechanical device needs to be in proper timing to function properly. If it is not, then the cam of the part of the system involved is either out of whack or adjustment some-how.

Stroke

A mechanical device needs to be have a proper stroke length to function properly. If it is not, then the cam follower of the part of the system involved is either out of whack or adjustment some-how.

Other

A mechanical device needs to be in proper position, timing, and stroke to function properly. If it is, but the machine is still not working right then something else in the system is involved or is out of whack or adjustment some-how (like a missing spring or a bent guide). Use visual inspection or feel to solve these problems. Turn to and do Exercise #1 and 2.

Chapter 13
Examples In The Rigging World

This profession is one that uses one, two, and three dimensional distance measurements (length, area, and volume) along with weight and force to allow you to move things safely.

Length

When estimating the weight of a load the first measurement that you take is that of the length of the object to be moved. This is usually done all in feet (or meters).

Area

When estimating the weight of a load the second measurement that you take is that of the area (the length times the width) of the object to be moved. This is usually done all in feet (or in meters).

Illustration 13-1

top view

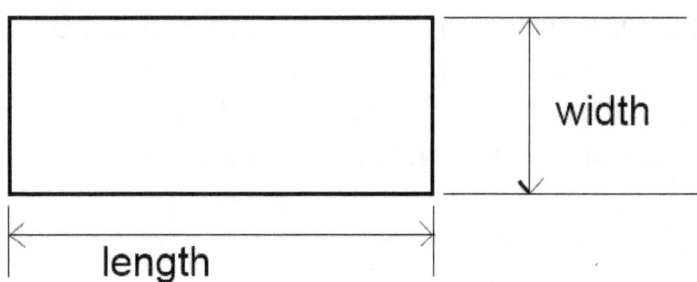

length

Volume

When estimating the weight of a load the third measurement that you take is that of the volume (length times width times height) of the object to be moved. This is actually the crushed height or the pancaked height of the object, not the measured height.

measured height   vs.   pancaked height

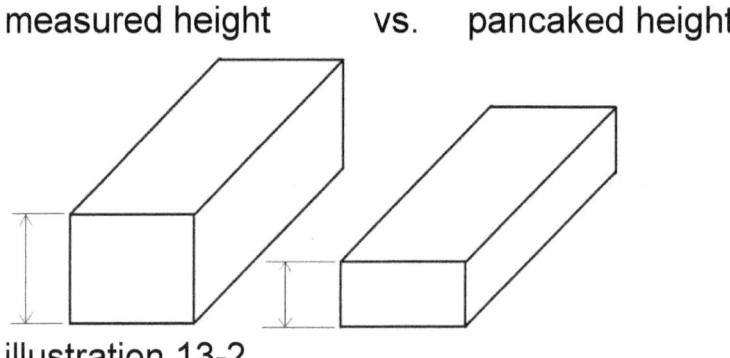

illustration 13-2

The actual measured height of an object would include large quantities of air. We want to measure the true length times the true width (which is relatively easy to do with a tape measure of some

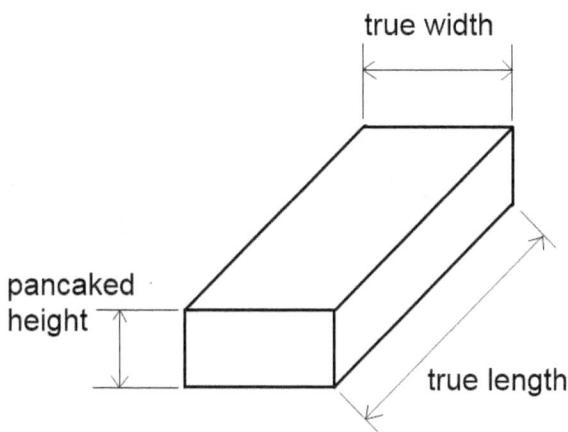

illustration 13-3

type), times the pancaked height of the object. This would ascertain the relative size in cubic feet.

This is important because it turns out that in the rigging world you can safely move objects using estimation because of the 300% safety factor involved in rigging equipment such as slings, chains, ropes, pry bars, etc. (50% for forklifts, cranes, and trucks).

Density

The rigging world can be thought of to consist of three sets of material weights (the three worldly substances) for estimating purposes:

Wet wood or water at about 50 lbs. {about 220 Nt.} per cubic ft. {.03 m³}

Stone or concrete at about 150 lbs. {about 675 Nt.} per cu. ft. {.03 m³}

Steel at about 500-550 lbs. {about 2400 Nt.} per cu. ft. {.03 m³}

wet wood or water	stone or concrete	steel

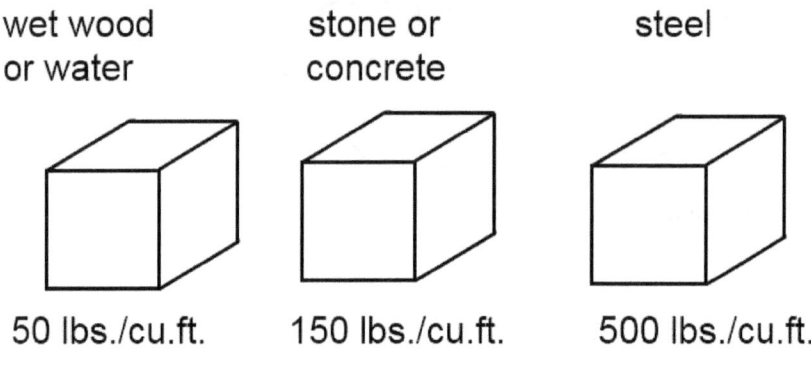

50 lbs./cu.ft.	150 lbs./cu.ft.	500 lbs./cu.ft.

Illustration 13-4

Weight

The weight of the object to be moved is therefore the measured length of the object, times the measured width of the object, times the pancaked height of the object, times the density of the object, or

L x W x PH x D. This gives you the estimated weight of the object to be moved.

If for example:

W=3 ft.   {1 m.}
L=10 ft.   {3 m.}
PH=1 ft.   {.3 m.}
D=50 lbs./cu.ft.   {220 Nt./.03 m³}

then weight = L x W x PH x D
w =10ft. x 3ft. x 1ft. x 50 lbs./cu.ft.
{3m. x 1m. x .3m. x 220Nt./.03m^3}
w = 1500 lbs.   {6600 Nt.}

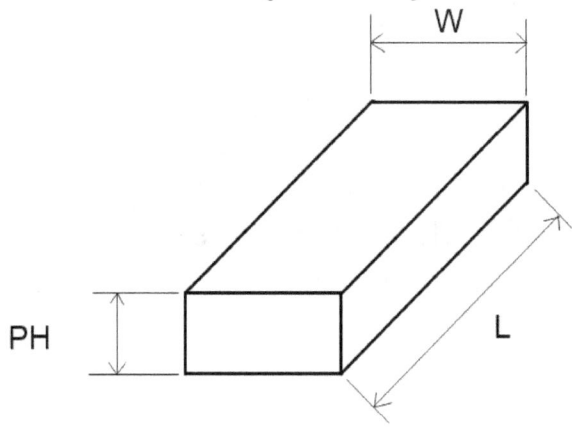

illustration 13-5

It is not important in most instances in the rigging profession for this to be exact, such as the weight of an object from a shipping document, because most rigging equipment has a built in safety factor of 3.

Working load limit (wll)
The current term used for safe working load (swl). It is the weight that a piece of rigging equipment can move safely with no shock loading.

Safe working load (swl)
The weight that a piece of rigging equipment can move safely with no shock loading. This term is generally being replaced by the term working load limit (wll) due to insurance industry concerns.

Breaking strength (BS)

This is the weight at which a piece of rigging equipment will fail or break.

Safety factor

For most rigging equipment except for vehicles, trucks, and cranes this is equal to 3. The relationship of this quantity in rigging is:

safe working load (swl) times 3 (the safety factor) is equal to breaking strength (BS)

or

swl x 3 = BS

or

BS/3 = swl

What this means as a practical matter is that your rigging equipment in general will not break unless you overload it by a factor of 3 or three hundred per cent (300%).

Manila rope

The swl of manila rope in tons is equal to the diameter of the rope squared or

swl = D(D)

or

swl = D x D   illustration 13-6

manila rope

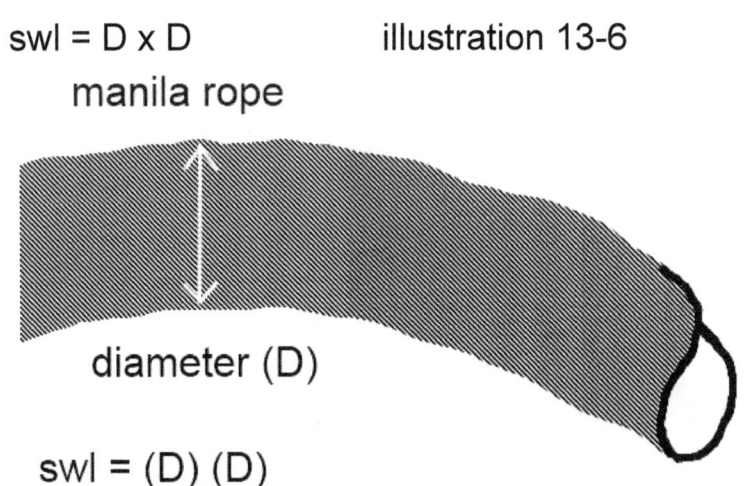

diameter (D)

swl = (D) (D)

Other fiber and artificial ropes
The swl of other ropes are adjusted from manila:

polypropylene +150% (the infamous yellow artificial hardware store staple)

polyethylene +150% (the artificial ski rope that feels like it has a sheen of oil on it)

nylon +200% (the shock loaders special)

sisal -35% (the other natural rope widely sold in hardware stores)

cotton -55% (the clothes line rope)
You remember clothes lines don't you?

Wire rope
The swl of wire rope in tons is equal to the 8 times the diameter of the wire rope squared or

swl = 8(D)(D)
or

swl = 8 x D x D   illustration 13-7

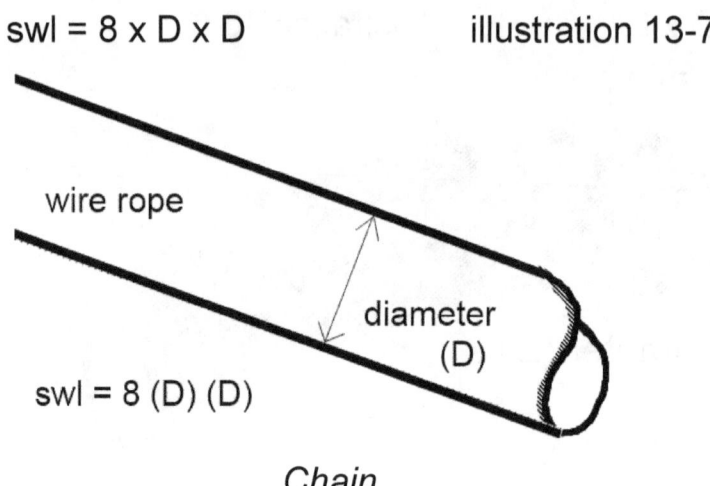

wire rope

swl = 8 (D) (D)

diameter
(D)

Chain

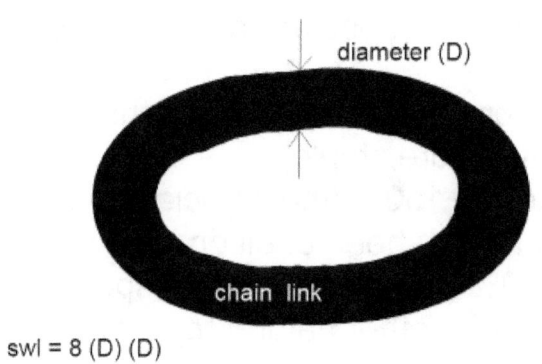

diameter (D)

chain link

swl = 8 (D) (D)

illustration 13-8

      The swl of chain in tons is equal to 8 times the diameter of the chain link squared or
swl = 8(D)(D)
or
swl = 8 x D x D

Shackles

The swl of shackles in tons is equal to the 8 times the diameter of the shackle pin squared or

swl = 8(D)(D)

or

swl = 8 x D x D

Illustration 13-9

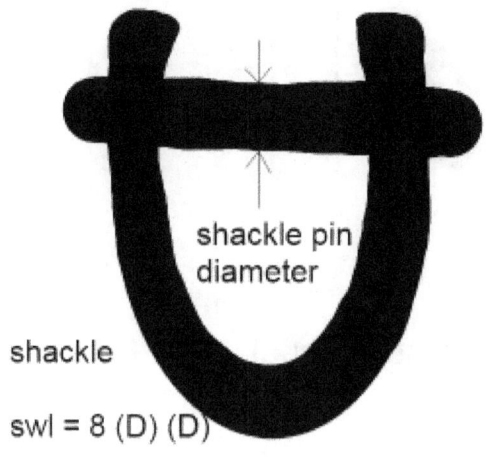

shackle pin diameter

shackle

swl = 8 (D) (D)

Hook

side view   front view

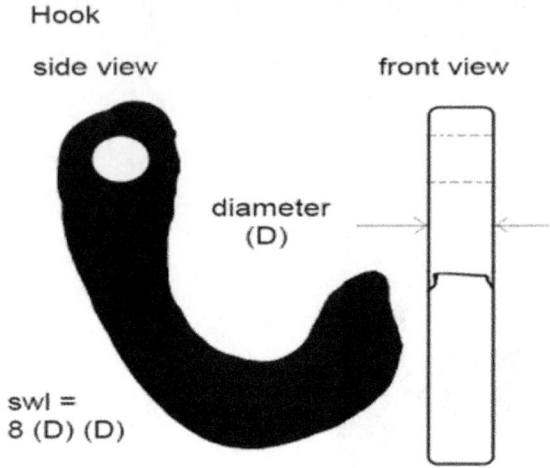

diameter
(D)

swl =
8 (D) (D)

The swl of hooks in tons is equal to the 8 times
the diameter of the hook's throat.

swl = the smallest throat measurement squared
swl = $8D^2$ or swl = 8 x D x D

illustration 13-10

Hand trucks
The swl of hand trucks in pounds is in the
range of 100 to 400 lbs. {440 – 1780 Nt.}.

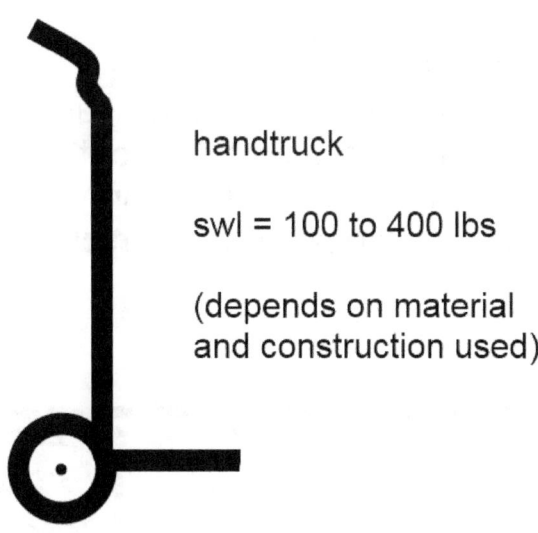

handtruck

swl = 100 to 400 lbs

(depends on material
and construction used)

illustration 13-11

Pry trucks
The swl of pry trucks is in the range of 1100
to 1800 lbs. {4900 – 8000 Nt.}.

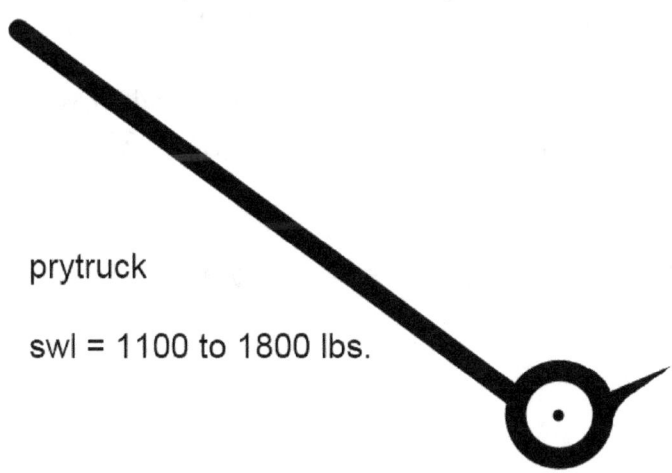

prytruck

swl = 1100 to 1800 lbs.

illustration 13-12

Levers

The weight that a lever can move is figured by taking the MA (mechanical advantage) or the long arm of the lever divided by the short arm of the lever by the weight of the person or persons using the lever.

lever

long arm   short
                                                  arm

Illustration 13-13

(for example, the long arm is 33" {840mm} and the short arm is 1 ½" {38mm})

$$MA = \text{long arm} / \text{short arm}$$
$$MA = 33"/1\ \tfrac{1}{2}" \qquad \{840mm/38mm\}$$
$$MA = 22$$

A 170 lb. {756 Nt.} person can move 3740 lbs. {16,640 Nt.} with this lever (22 x 170# = 3740#) {22 x 756 Nt. = 16,640 Nt.}.

Looking at a pry truck lever example:

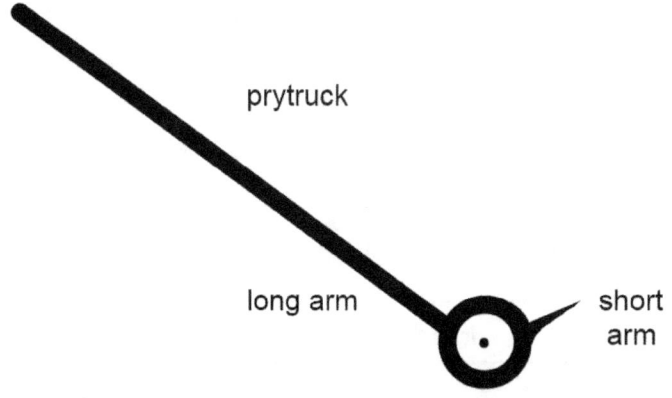

prytruck

long arm · short arm

illustration 13-14

(for example: the long arm is 78" {1980mm} and the short arm is 6 " {152mm})

MA = long arm / short arm

MA = 78"/6"   {1980mm/152mm}

or MA = 13

A 170 lb. {756 Nt.} person can move 2210 lbs. {9830 Nt.} with this pry truck (13 x 170# = 2210#) {13 x 756 Nt. = 9830 Nt.}.

Slings

The weight that a sling can move is dependent on its rated capacity (listed on a tag on the sling itself) times the sin of the angle that the sling makes with the load times the # of legs or

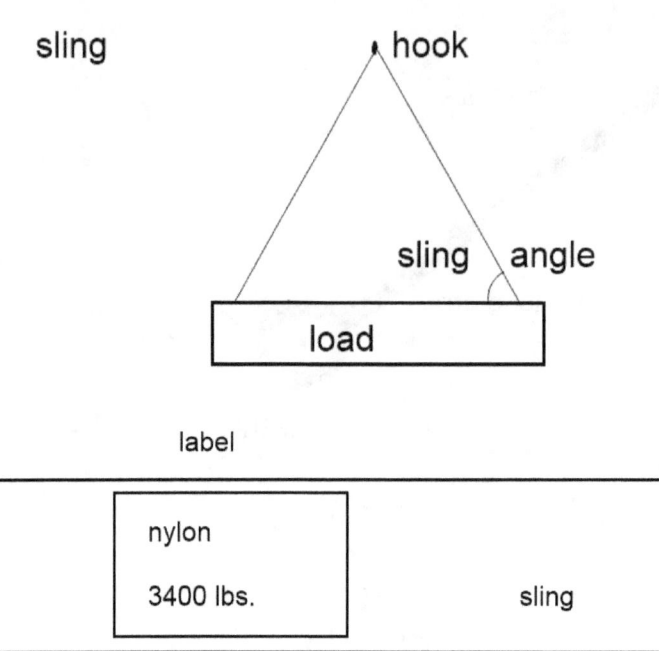

sling   hook

sling   angle

load

label

nylon

3400 lbs.   sling

{15,130 Nt. sling capacity}

Illustration 13-15

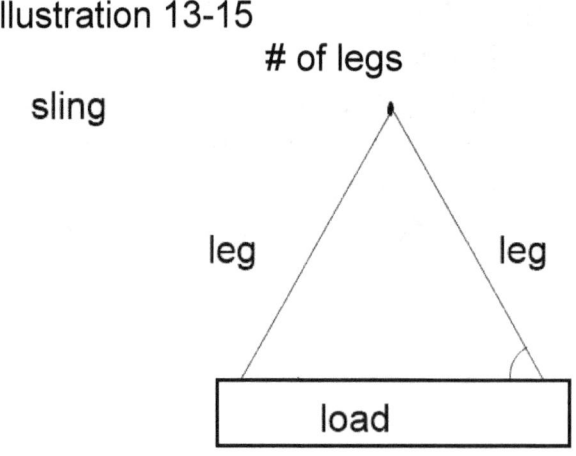

# of legs

sling

leg   leg

load

2 legs on this lift example

illustration 11-16

Formula example:

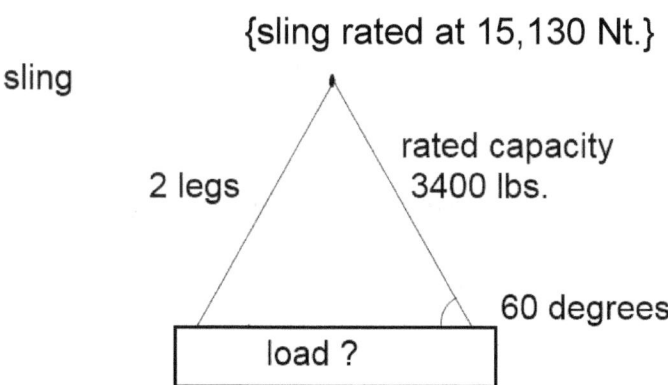

{sling rated at 15,130 Nt.}

sling

2 legs

rated capacity
3400 lbs.

60 degrees

load ?

illustration 13-17

sling x (sin of the angle) x (# of legs) = swl

(3400 lbs.)(sin 60 degrees)(2 legs) = swl
{(15,130 Nt.)(sin 60 degrees)(2 legs) = swl}

(3400)(.866)(2) = swl   {15,130(.866)(2)}

5888 = swl   {26,205 = swl}

swl = about 5900 lbs. = {about 26,000 Nt.}
        Please take the following rigging practice
test. The answers are in appendix B. Hints are in
[brackets].

(1) What knot makes a temporary eye?

(2) What knot is used to make a temporary
handrail? [on a construction site]

(3) What knot is used to cinch a rope on an object? [boating]

(4) What are fiber rope thimbles made of?

(5) Is it safe? {weight = 380,000 Nt.}

sling

bridle
2 legs

sling material 1 1/2"
wire rope

45 degrees

weight 85,500 lbs

illustration 13-18

sling material is 1.5" {38mm} dia. wire rope
hookup is a bridle (2 legs)
sling angle is 45 degrees
weight of load is 85,500 lbs. {380,000 Nt.}

(6) What are the first two over riding principles of moving loads in rigging?

(7) The world generally consists of three materials. What are they and what do they weigh?

(8) The concrete floor can usually take how much stress?

[in PSI]

(9) How much pressure will the pry bar develop with two people pulling on it?

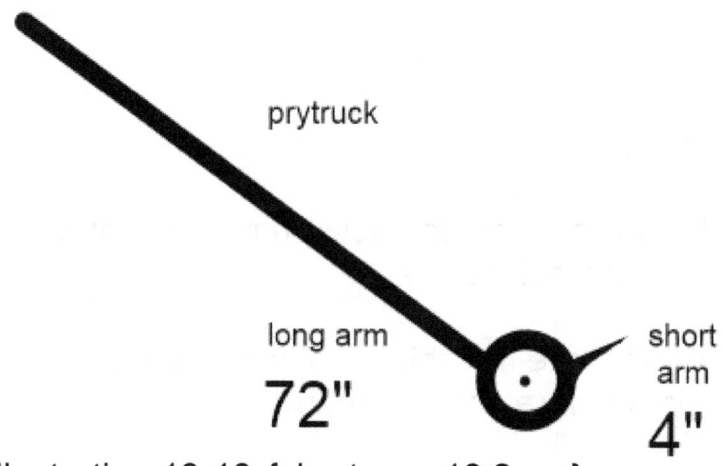

prytruck

long arm

72"

short arm

4"

illustration 13-19 {short arm 10.2 cm}

long arm is {1.83 m.}, point of contact is 2 square inches {13 cm²}

(10) The two types of rigging hooks are called what?

(11) What does a mouse do?

(12) If you use wire clips to make an eye in a wire rope the swl decreases by about how much?

(13) What is the rule of thumb between swl and BS?

(14) If you were made up of 50% concrete, then how much would you weigh?

(15) Whipping refers to what?

(16) What is a tag line?

(17) What is cribbing?

(18) Extension ladders should have how much overlap when extended?

(19) What is the first thing that you should do to all ladders before you use them?

(20) Is it safe why or why not?

 sling

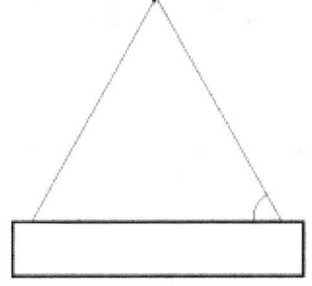

sling is nylon with a 3800 lb. {16,900 Nt.} capacity
sling hookup is a double basket (4 legs)
sling angle is 60 degrees
weight of load is 6800 lbs. {30,260 Nt.}

illustration 13-20

Turn to and do Exercise #3 and 4.

Chapter 14
Examples In The Fluid Power World

This area, like the electrical area that follows uses a lot of four dimensional concepts (pressure & flow) and eight dimensional concepts and terms (power) to describe what occurs in hydraulics and pneumatics along with six dimensional ones (force) and seven dimensional ones (torque & energy).

Pressure

This is usually defined as the weight of a column of material above the surface of interest and is usually given in pounds per square inch (PSI) {kPa}, atmospheres (atm.) [bar], or newtons per square meter {Pascal}. The most familiar point on the pressure scale is sea dimension pressure which is about 14.7 PSIG {101.3 kPa} or 1 bar.

Pressure Scale

In order to have a scale of any type you need to know the units involved and at least two points on the scale defined. The four most common scales in fluid power are the gage pressure scale, the absolute pressure scale, the vacuum pressure scale and the atmospheric pressure scale.

Head

A concept that is often caught up with pressure scales is the concept of head. A head is the equivalent height of a particular material in linear measure (ft., in., m., etc.) above a surface area of interest (usually in square inches) that gives equal pressure. We have the approximate situation that,

30 ft. {9 meters} of water,
30" {.75 m.} of Hg, and
30 miles {48 km} of the Earth's atm. are all equal to about 15 PSI {100 kPa}.

Flow

In the hydraulics area, fluid power flow is usually given in gallons per minute (GPM) {LPM}. Flow most directly affects the speed of the components in the circuit such as the extension and retraction of cylinders,

illustration 14-1

double acting cylinder

and the rotation (RPM) of hydraulic motors.

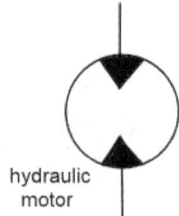

hydraulic motor

illustration 14-2

In the pneumatics area, fluid power flow is usually given in standard cubic feet of air per unit time (SCFM or SCFH) {LPM or LPH}. Flow most directly affects the speed of the components in the circuit such as the extension and retraction of cylinders,

single acting cylinder

and the rotation (RPM) of air motors.

illustration 14-4

HVAC

In the HVAC area (heating, ventilating, and air conditioning) air movement is also given in SCFM (standard cubic feet per minute) {LPM liters per minute}. This would apply to ventilating fans in open systems such as those mounted overhead and for those in closed systems such as those in ductwork.

Pressure Effects

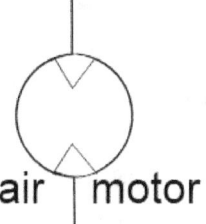

air motor

illustration 14-5

In the hydraulics and pneumatics areas of fluid power, pressure is usually given in pounds per square inch (PSI) {kPa}. Pressure most directly affects the output forces or output torques of components in the circuit. Force is associated with the extension and retraction of cylinders. Torque (ft.lbs.) {Nm} is associated with the rotation of hydraulic and pneumatic motors.

Power

In the fluid power area power is usually given in units of horsepower (HP) or in units of kilowatts [kW].

Turn to and do Exercise #5 and 6.

Chapter 15
Examples In The E&M World

This area generally uses four dimensional concepts (voltage, current) through eight dimensional concepts and terms (power), zero dimensional terms (such as resistance) and minus one dimensional concepts (inductance and capacitance) in order to better understand the what and the how of the workings of electricity.

Voltage

This is defined as electrical pressure. It works just like the pressure in your municipal water system except you are pressing on electrons instead of water particles. Electrical pressure or voltage will usually not kill humans. The units of electrical pressure is volts.

Current

This is defined as a flow of charge. It works just like the flow in your water pipes at home except you are causing electrons to move around in a path instead of water in a pipe. Current is what will harm or kill you in an electrical circuit if you are not careful. The various electrical thresholds in a typical office or home situation are illustrated below.

illustration 15-1

15 ma. 150 ma. 15 amps

too much
for your
heart

too much
for your
muscle

too much
for your
flesh

The units of electrical flow are amps. The question often comes up about what causes the most problems in a circuit. Well, look at the water situation around your home. A typical system might have 40 PSI {275.8 kPa} of pressure. That pressure will be the same at your garden house, at the fire hydrant, and at the main water line coming into your housing development. Now, suppose that you break into a water line or turn on a water line at each of these three places. What would happen?

Well, at the garden hose in your backyard everyone would be mad because they all just got soaking wet. At the fire hydrant near your front yard you would be knocking over everyone in the neighborhood as they were hit by the fire hose water stream. At the entrance to your neighborhood the break in the water main would probably be creating some sort of a monstrous eroded hole that was attempting to swallow up a few cars.

The important thing to remember in all of this is that in all of these instances is that the pressure was exactly the same (40 PSI) {275.8 kPa} the only thing that changed was the current flow (GPM).

Resistance

The resistance to flow is defined in electricity in Ohms law and has the units of ohms (omega).

Ohms law

The most general relationship that is learned in intro. courses in electricity (for direct current and resistive alternating
current circuits) is Ohms law. It states that the voltage (E or V), is equal to the current (I) times the resistance (R).

$$E = I\,R \text{ or } V = I\,R$$

Power

When determining the amount of energy used in a given circuit in a certain amount of time, you would use power. The units of power are

watts. Where electrically, 1 horsepower = 746 watts and where mechanically,
1 horsepower = 550 ft. lbs./sec. {.75 kW}

Power Law

The most helpful relationships next to Ohms law and the horsepower conversions listed above are the two main power laws. The first comes from the general fluid mechanics relationship that says that the power (P) is equal to flow times pressure.

$$P = (flow)(pressure)$$

In electrical work it is used in this fashion,

$$P = I V$$

where P = power in watts, I = current in amps, and V = voltage in volts. In the field it is also stated as, watts equals amps times volts, or watts = (amps)(volts).

The other power relationship that is used in the electrical area quite a bit is the one that states that power is equal to the current squared times the resistance, or P equal I squared R.

$$P = (I^2)R$$

This is used for example when you need to measure the resistance of a heater element in ohms with a known voltage source and you wish to calculate the kilowatts (KW) of a replacement heater.

Series Resistance

Let's look at a typical circuit situation with resistors hooked in series. If you had three resistors like this,

10   100   190

r1   r2   r3

illustration 15-2
hooked up to a 24 volt battery what would be the voltage drop across each resistor? The two facts that will help you solve this problem are that the total voltage drop in the circuit (across the resistors) must be equal to the total voltage rise (the battery) in the circuit.

You are given the three resistor values so you would add them together to get the total circuit resistance (R_t). You know that the total circuit voltage (V_t) is equal to the voltage of the battery. So using Ohms law to solve for the total circuit current (I_t).

$$V_t = (I_t)(R_t)$$

substituting

$$24 = (I_t)(10 + 100 + 290)$$

$$24 = (I_t)(400)$$

$$24/400 = I_t$$

$$I_t = 24/400$$

$$I_t = .06 \text{ amps or 60 ma.}$$

Once you have the total circuit current (I_t) you can solve for the voltage drop at each resistor using Ohms law, as the same amount of current will be flowing through each resistor as they are hooked in series.

$V1 = (I_t)(R1)$ $\qquad\qquad$ $V2 = (I_t)(R2)$

$V3 = (I_t)(R3)$

substituting

$V1 = (.06)(10)$ $\qquad\qquad$ $V2 = (.06)(100)$

$V3 = (.06)(290)$

we have,

$V1 = .6$ volts $\qquad\qquad$ $V2 = 6$ volts

$V3 = 17.4$ volts

To double check your answer just add up the three voltage drops and they should equal the voltage rise of the battery (24 volts in this example).

Parallel Resistance

Let's look at a typical circuit situation with resistors hooked in parallel. If you had three resistors like this,

illustration 15-3

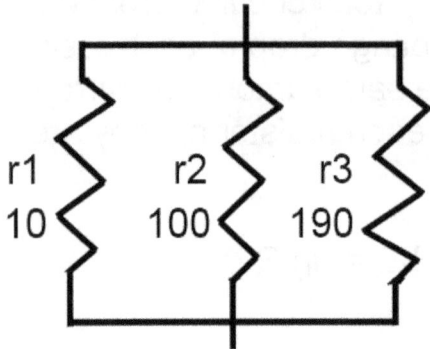

r1 10   r2 100   r3 190

hooked up to a 24 volt battery, then what would be the voltage drop across each resistor and the current going through each resistor? There are two facts that will help you solve this problem. The sum of the currents in the circuit (through each resistor) must be equal to the total current (through the battery) in the circuit. The voltage drop across each resistor hooked in parallel is the same.

The voltage drop across each resistor is equal to the voltage rise of the battery (in this example 24 volts).

$V_t = V1 = V2 = V3 = 24$ volts

You are given the three resistor values so you would add them together as reciprocals to get the total circuit resistance (R_t).

$$R_t = \frac{1}{1/R1 + 1/R2 + 1/R3}$$

$$R_t = \cfrac{1}{1/10 + 1/100 + 1/290}$$

$$R_t = \cfrac{1}{.1 + .01 + .00345}$$

$$R_t = \cfrac{1}{.11345}$$

$R_t = 8.815$ ohms

     You can double check your answer in the field by the fact that the total circuit resistance of any number of resistors hooked in parallel will always be less then the value of the smallest resistor in the parallel network. In this case, is 8.815 ohms less than 10 ohms? Yes it is.

     You know that the total circuit voltage (V_t) is equal to the voltage drop across each resistor in the parallel network. So use
Ohms law to solve for the total resistor network circuit current flow (I_t).

$V_t = (I_t)(R_t)$

substituting

$24 = (I_t)(8.815)$

$24/8.815 = I_t$

$I_t = 24/8.815$

$I_t = 2.723$ amps

Once you have the total circuit current (I_t) you can solve for the current flow through each resistor using Ohms law, as the same amount of voltage will be across each resistor, as they are hooked in parallel.

$V1 = (I1)(R1)$ $V2 = (I2)(R2)$

$V3 = (I3)(R3)$

substituting

$Vt = (I1)(10)$ $Vt = (I2)(100)$

$Vt = (I3)(290)$

$24 = (I1)(10)$ $24 = (I2)(100)$

$24 = (I3)(290)$

$24/10 = (I1)$ $24/100 = (I2)$

$24/290 = (I3)$

$2.4 = (I1)$ $.24 = (I2)$

$.083 = (I3)$

(I1) = 2.4 amps   (I2) = .24 amps =240
                             ma.
   (I3) = .083 amps = 83 ma.

         To double check your answer just add up
the three current flows and they should be equal
the current flow of the battery (2.723 amps in this
example).

Turn to and do Exercise #7 and 8.

Chapter 16: Exercises

Exercise #1
              Approximation and walking
         Walk around the perimeter of the building
you are in. What is the distance in feet? What is
the distance in meters?
Remember that two paces is about equal to 5 feet
and that four paces is about equal to 3 meters (10
ft.).
              Exact measurement
         Take a tape measure and walk around the
perimeter of the building that you are in and
measure its perimeter in feet or meters.
              Unit comparisons
         Compare your approximate measurement of
the perimeter of the building with your exact
measurement. Give your answer as a per cent,
as a decimal number, as a ratio, and as a fraction.

Exercise #2
>Area
>You are involved in the selling of a house. It is 40 ft. {12.2 m.} long and 28 ft. {8.5 m.} wide. It is two stories with an unheated full basement. What is the total heated square feet living area of this house?
>Conversions   The site that the above mentioned house is sitting on is out in the country on a rectangular lot that measures 117 ft. {35.7 m.} by 390 ft. {119 m.} How many acres is the lot? 1 acre is about equal to 208 ft. times 208 ft. {63.4 m. x 63.4 m.}
>Speed and velocity
>An automobile is traveling down the road at 60 fps. {18.3 mps}. How many mph {kph} is that?
>1 mile = 5280 ft. = {1.6 km.}
>A car drives through a town going 40 mph {64.4 kph} for 30 minutes, and then travels further north at 60 mph {96.6 kph} on an interstate highway for another hour and a half. What is the average speed of the car? What is the average velocity of the car? How many miles has the car traveled in 2 hours?

Exercise #3
>Volume
>A rectangular tank measures 3 ft. by 6 ft. by 4 ft. {.91 m. x 1.83 m. x 1.22 m.} How many cubic feet {cubic meters} are in the tank? How many gallons of water would fit in the tank? How many

liters of water would fit in the tank? 1 cubic ft. = 1728 cu. in., 1 gallon = 231 cu. in., and 1 cubic meter = 1000 liters = 1057 quarts.

A slab for the floor of a building is 40 ft. wide by 60 ft. long by 6 inches thick. {12.19 m. x 18.23 m. x 152mm}. How many cubic yards {cubic meters} of concrete do you need to order for this project?

Acceleration in your seat.

Your neighbor's car can go zero to sixty in 11 seconds. Assuming a steady acceleration, what is the value of this acceleration in feet per second squared {mps^2}? 1 mile = 5280 ft. = {1609 m.}

Gravity and research

On the Earth you would fall 16 feet {4.88 meters} in one second. How far would you fall on the moon in one second?

Mass and metric

The machine that was shipped to your company has a mass of 1800 kilos. What would your mass be in kilos?

Charge a joke

A dust devil may have up to 4000 volts per foot inside of it. Is this a bigger charge then the state of your credit card after a day of shopping?

Exercise #4

GPM

A hydraulic system on a machine has a volume flow of 10 gallons per minute (10 GPM)

{37.85 LPM}. If the cylinder on the machine has a bore size of 3 inches {76.2mm} and a stroke of 18 inches {457mm}, how long will it take to extend or retract? where1 gallon = 231 cu. in.

Standard cubic feet

A compressor system in a plant has a volume flow of .04 SCFS {.00113 m³}. If the size of the receiver next to the compressor is 28 cu. ft. {.793 m³}, then about how long will it take to fill it up to system pressure?

Airplanes

A plane burns 8 kg. of fuel per minute. How much mass flow is that per second? How many hours can the plane fly with a 45 minute fuel reserve if the tank holds 2500 kg.?   Electrons

A 150 amp branch circuit in a building has a short circuit of 40,000 amps for 1/2 cycle before a circuit breaker trips. How many coulombs of charge (Q) were involved in the short circuit? 1 cycle = 60 Hz.

Visualization

A line is bounded by 2 points. A square is bounded by 4 lines. A cube is bounded by 6 squares and a Tesseract is bounded by 8 cubes. What is a circle and a sphere bounded by?

Exercise #5

Humdinger

A 5 kg. sledge hammer hits a Hummer going from 0 to 20 kph in 1/2 sec. A big dent and an argument ensue. What was the momentum of the sledgehammer on impact with the Hummer?

Basically bullets

A bullet with the same momentum as the sledge hammer in the problem above, with a mass of .01 kg. strikes a target.

How fast was the bullet going?

Exercise # 6

In the way

Someone who was perturbed pushed a 900 lb. {4005 nt.} box of machine tool parts down the aisle for about 10 1/2 ft. {3.2 m.} ending up just short of a wall. How much did they weigh?

In the way again

Someone who was perturbed pushed a 400 kg. box of machine tool parts down the aisle for about 3 meters ending up just short of a wall. How much force did they use if they did it in 3 seconds?

Exercise #7

In the way again, revisited.

Someone who was perturbed pushed a 900 lb. {4005 nt.} box of machine tool parts down the aisle for about 10 1/2 ft. {3.2 m.} ending up just short of a wall. How much work did they do?

Quitter

My 8500 BTU {2.49 kW} air conditioner quit. How much energy in joules and/or calories are we talking about?

Tightened up.

Someone tries to tighten up a stuck nut on a 1/2 inch {12.7mm} diameter bolt on their truck. They are using a wrench with an 18 inch {457mm} handle on it. It is still stuck. A 4 foot {1.2 m.} pipe is lying nearby. If they slide the 4 foot {1.2 m.} pipe over the handle of the wrench and pulled, what torque was developed just before the bolt broke off?

Future fuels internet lookup.

How many calories are available in the typical jelly donut?

Exercise #8

Up and at 'em.

13 people cram into an elevator that takes 10 seconds to get to the third floor. How much energy is needed if each floor is 14 ft. {4.3 m.} tall and each person averages 170 lbs. {756 nt.}?

Horses of course.

A 20 HP {894,600 joules/min.} motor is running at 1740 RPM. If you attach a 10 lb. {44.4 nt.} weight distributed evenly 1 foot {305mm} from the center of the rotating shaft, what would the kW of the motor be?

Appendix -a-

A

acceleration

a

$L/(T)(T)$

dv/dt

ft./sec.^{2}

m./sec.^{2}

acceleration of gravity = g

32.2 ft./sec.^{2} = g

9.81 m./sec.^{2} = g

acre = 43,560 sq.ft.

amp = a unit of current

Q/T

ampere = current flow

1 amp = 1 coulomb/sec.

AND gate

    series

angular acceleration

$$\alpha$$

$$1/T^2$$

angular displacement

$$\theta$$

angular velocity

$$\omega$$

$$1/T$$

angular momentum

L

ML^2/T

B

BTU = a unit of energy

British thermal unit

1 BTU = 252 calories

288,000 BTU = 1 ton cooling or heating capacity in HVAC

(that translates into 12,000 BTUs per hour)

buffer

YES gate

C

calorie = a unit of energy

252 calories = 1 BTU

1000 calories = 4186 joules

capacitance = $\mathbb{C}$
$\quad$ Q^2T^2/ML^2

charge = Q

charge flow = Q/T

capacitance

$\quad$ 1 farad = 1 coulomb/volt

circle

circular area = $A = \pi r^2$

coulomb = (current in amps) (time in seconds)

$1.6(10^{-19})$ coulomb = 1 electron charge

current

amp

Q/T

current density = $\mathbf{J}$

$Q/T \ L^2$

$amp/meter^2$

D

degree Fahrenheit

    freezing H_2O = 32 degrees F

    boiling H_2O = 212 degrees F

degree Rankin

    freezing H_2O = 491 degrees R

    boiling H_2O = 671 degrees R

degree Celsius

> freezing H_2O = 0 degrees C

> boiling H_2O = 100 degrees C

degree Kelvin

> freezing H_2O = 273 degrees K

> boiling H_2O = 373 degrees K

density

> M/L^3

> M/volume

> kg/m^3

displacement

> L or s

E

electric dipole moment = $\mathbf{P}$

QL

coulomb-meter

electric displacement field = $\mathbf{D}$

Q/L^2

coulomb/meter^2

electric flow = voltage

electric field = $\mathbf{E}$

ML/QT^2

volts/meter

electric flux = Φ_E

$$ML^3/QT^2$$

volt-meter

electric pressure = current

electron charge = $1.6(10^{-19})$ coulomb

energy

work

torque

energy units

joules

BTUs

calories

$$ML^2/T^2$$

ft.lbs. (foot-pounds)

oz.in. (ounce-inches)

nt.m. {newton-meters}

F

farad = coulomb/volt

flow

    mass flow

        M/T

        kg./sec.

        kg./min.

        kg./sec.

    volume flow

        L^3/T

        GPM (gallons per minute)

SCFM (standard cubic feet per minute)

m³/min. {cubic meters per minute}

SCFH (standard cubic feet per hour)

m³/hr. {cubic meters per hour}

current flow

Q/T

amps

force

ML/T^2

lbs. (pounds)

nt. {newtons}

F = ma (force = mass times acceleration)

foot = a unit of length

G

gravity

g

32.2 ft./sec.^{2}

9.81 meters/sec.^{2}

gravitational acceleration

L/T^2

H

hertz = a unit of frequency

Hz.

60 Hz in the U.S.

50 Hz. in Europe

horsepower

HP

mechanical horsepower

1 HP = 550 ft.lbs./sec. = 746 nt.m./sec.

electrical horsepower

1 HP = 746 watts

hysteresis (known as dead band in process control and HVAC)

I

induction = **L**

ML^2/Q^2

henry

1 henry = 1 volt(sec.)/amp

J

joule = a unit of energy

1055 joules = 1 BTU

4.186 joules = 1 calorie

1.356 joules = 1 ft.lb.

K

kilogram = a unit of mass

kg.

M

mass

kWh (kilowatt-hour)

L

length

L

foot (12 inches)

inch (1000 mils)

yard (36 inches)

mile (5,280 ft.)

meter

milli-meter {1/1000 meter}

kilometer {1000 meters}

linear (straight line motion)

M

magnetic dipole moment = μ_m or $\mathbf{M}$

QL^2/T

amp-meter^2

magnetizing field = $\mathbf{H}$

Q/TL

amp/meter

magnetic field = $\mathbf{B}$

M/QT

tesla

weber/L^2

magnetic flux = Φ_B

ML^2/QT

weber

volt-sec.

mass

M

kilogram

kg.

meter = a unit of length

micron = 10^{-6} meter

Mks {meters, kilograms, seconds}

momentum

p

M(L)/T or ML/T

p = mv

N

newton

nt.

weight

force

NOT gate

inverter

O

ohm = a unit of resistance

ohm's law

E = IR

voltage = (current) (resistance)

volts = (amps) (ohms)

pressure = (flow) (resistance)

OR gate

parallel

P

parallel

OR gate

period

1/T

1/frequency

phase

single phase (residential)

three phase (industrial)

pressure

atmosphere (atm.)

bar

inches of mercury ("Hg.)

newton per square meter {nt./m^2}

pounds per square inch (PSI)

head

approx.   **more exact**

30 ft. H$_2$O = 15 PSI
        33.9 ft. H$_2$O = 14.7 PSI

{9 m. H$_2$O = 100 kPa
        9.14 m. H$_2$O = 101.4 kPa}

30 in. Hg. = 15 PSI
29.92 "Hg. = 14.7 PSI

{3/4 m. Hg. = 100 kPa
.762 m. Hg. = 101.4 kPa}

30 mi. Earth's air = 15 PSI
1 atm. = 14.7 PSI

{48 km. Earth's air = 100 kPa
1 atm. = 101.4 kPa}

100,000 nt./m^2 = 15 PSI
101,300 nt./m^2 = 14.7 PSI

Q

Q (charge)

   coulomb

   1 coulomb = (1 volt) (1 farad)

R

resistance

> electrical resistance

> ohm

> mechanical resistance

> friction

S

sink

> conventional current (-)

source

> conventional current (+)

speed

L/T

dl/dt

feet per second (fps)

miles per hour (mph)

meters per second {mps}

kilometers per hour {kps}

T

temperature

    standard

        Fahrenheit

        Rankin

    metric

Celsius

Kelvin

time

t

T

ton (HVAC) – 1 ton of cooling or heating in HVAC = 288,000 BTUs

torque

ML^2/T^2

ft.lbs

oz.in.

nt.m.

U

unity = 1

units

    standard

        feet

        slugs (mass)

        seconds

    metric {Mks}

        meters

        kilograms

        seconds

V

velocity (magnitude and direction)

L/T

volt = unit of electrical voltage

voltage

electrical pressure

ML^2/QT^2

W

watt = power

746 watts = 1 HP

1000 watts = 1 kW

weight

ML/T^2

pounds

newtons

force

work

W

ft.lbs.

energy

x-ray

YES gate

buffer

Z

Zero

Appendix -b-: Selected Answers

Chapter 13 Rigging test
1) the bowline

This knot is also used in search and rescue work to make loops in ropes for arms and legs that will not close down on a person and cut off circulation.

2) the clove hitch

Nobody wants folks pitching off the unfinished floor of a building.

3) two half hitches

This knot creates a loop that does choke down on the object that you tie it on.

5) no

swl of sling = 36,000 lbs. {160,200 nt.} [swl = 8 D squared in tons]

times 2 legs = 72,000 lbs. {320,400 nt.}

times the sine of the angle (.707) = 50,904 lbs. {226,500 nt.} [the sine of 45 degrees is .707]

The load weighs 85,500 lbs. {380,000 nt.} but the sling configuration will only support about 51,000 lbs. {226,500 nt.} safely.

6)   A. Never put your hand, your foot, or yourself directly under the load, because one day it will fall.
      B. Move the load as slow as you think that you need to go, and then move it three times slower.

7)   wet wood or water at around 50 lbs./cu. ft. {220 nt./.03 m³}

      stone or concrete at around 150 lbs./cu. ft. {660 nt./.03 m³}

      steel at around 500 lbs./cu. ft. {2200 nt./.03 m³}

8) usually in the neighborhood of 9000-11,000 PSI {62,000 – 76,000 kPa}

9) 3060 PSI {21,100 kPa}

MA = 72 in./4 in. {1.83m/10.2 cm} = 18
[mechanical advantage]

force = 2 people times 170 lbs. {756 nt.} = 340 lbs.
{1512 nt.}
times MA of 18 = 6120 lbs. {27,200 nt.}

pressure = force times contact area = 6120 lbs. in
2 square in. = 6120 lbs. times 1/2 = 3060 PSI
{27,200 nt. x .5 = 13,600 nt./6.45 cm^2} = {21,100
kPa}

10) slip hook and grab hook

11)   It keeps the sling from coming out of the
throat of the lifting hook.

12) 15-35% depending on how well you made the
connection.
99

13) BS = 3 (swl)
The breaking strength is equal to about 3 times
the safe working load.

14) about 400 lbs. {1780 nt.}
If you were a big strapping person of say 200 lbs.
and since you are mostly made of water, you
would have a volume of around 4 cu. ft. {.113 m^3}.
2 cu.ft. {.057 m^3} of concrete = 300 lbs. {1335 nt.}

2 cu. ft. {.057 m³} of water = 100 lbs. {445 nt.}

15) What it is that you do to keep the strands on the end of a rope from unraveling.

16) An attached rope on the end of a suspended load that a person holds onto to prevent the load from pivoting around in unexpected ways is a tag line.

17) Cribbing is wood timbers used to support the bottom of a load temporarily.

18) 15% minimum

On a 20 ft. {6 meter} extension ladder this means that the minimum overlap would be 3 ft. {.9 meter}.

19) inspect them

20) yes

swl of sling = 3800 lbs. {16,900 nt.} [the swl is on the label of the nylon sling]

times 4 legs = 15,200 lbs. {67,600 nt}

times the sine of the angle (.866) = 13,163 lbs. {58,136 nt.} [sine of 60 degrees is .866]

The load weighs 6800 lbs. {30,300 nt.} and the sling configuration will support about 13,000 lbs. {58,000 nt.} safely.

Exercise #4 Visualization

circle - bounded by 1 line (the circumference) - $\pi(d)$ - $2\pi(r)$

sphere - bounded by 4 areas - $4\pi r^2$

4D hyper sphere - bounded by 6 spherical volumes - $8\pi(r^3)$

Exercise #6 In the way

about 170 lbs. {756 nt.}

Exercise #7 Quitter

Even though the air conditioner says 8500 BTU {8,968 kJ} it really is 8500 BTU/hr. {2.49 kW}.

Abstract:
Technical Physics
The Eleven Dimensions of
Contemporary Physics

Estimation

For me it was two major epiphanies. The first came about in a college Physics class. You know the type of classroom that you see in the movies, or you may have even experienced it yourself. The blackboards completely surround you on every side in the classroom. My professor, Dr. Antonakos was on the fourth board working his way around the room when he turns and looks at one of the boards in the front of the room and says, "Well, that (meaning that top half of the front board) is about equal to 1/3. Next we will substitute it into this equation here."

I do not know about you, but in all of my math classes through High School it was pretty much drilled into my head that you had to be exact or pretty darn near exact in all of the math equations and problems that you worked on. So, it was a pretty big light bulb going off in my head when I realized that you can estimate a lot of the time and come out pretty close to the "right" answers. I knew about the old saw that says, "Close only counts in horseshoes, hand grenades, and atomic warfare", but this was a big deal to me. Since that time I have used estimation quite frequently in the field in various endeavors as diverse as rigging and fluid mechanics.

Dimensions
 The second epiphany came about while walking along some railroad tracks one night. I had donated my car to the local community college a month or two previously, so that they could do repair work on it for practice in their automotive classes. I had been riding my bike to and from work. I had just been hit by a car a week or so previously (but not seriously injured). So, now I was riding the bus to work during the afternoon and walking home from work at night.
 As I was walking down the railroad tracks that night, I had come to realize that all of the equations in the engineering texts and in all of the Physics texts that I had encountered were in fact all interrelated to each other. I realized that instead of memorizing all of the equations, or looking up equations for all of the thousands of situations that one could or would run into, you could see them as a relationship instead. You would be able to derive equations and relationships from what you already knew in your head. The first way equations and concepts are related is that they all have M, L, T, and Q units.

dim. -2	dim. -1	dim. 0	dim. 1	dim. 2	dim. 3
μ_o					
			L		L^3
permeability		1		L^2	
	T		1/T		L/T^2
ML/Q^2		unity		L/T	
	henry				M

henry/meter	**L** induc-tance	L to the zero power		v	
	ML²/Q²				
ε_0		T to the zero power			
Q²T²/ML³	farad				
permittivity	**C** capaci-tance				
farad/meter	Q²T²/ML²				

dim. 4	dim. 5	dim. 6	dim. 7	dim. 8
	mv		Fs	
M/T		ma		IV
	p		W	
Q/T		F		P
	p = mv		W = Fs	
		F = ma		P=IV
	ML/T		ML²/T²	
		ML/T²		ML²/ T³

The second way that they were related was dimensionally. If one assigned the 1st dimension to the L unit, the 1st dimension to the 1/T unit and 3rd dimension to the M and the Q unit then L and 1/T would be in the dimension 1 area.

152

Dim -2	Dim -1	Dim. 0	Dim. 1	Dim. 2	Dim. 3	D4	D5	D6	D7	D8
			L							
		1								
			1/T							
		unity								
			length							
		L to the zero power								
			freq.							
		T to the zero power								
			Hz.							
		resis-tance								

L^2 and L/T would be in the dimension 2 area.

Dim -2	Dim -1	Dim. 0	Dim. 1	Dim. 2	Dim. 3	D4	D5	D6	D7	D8
			L							
		1		L^2						
			1/T							
		unity		L/T						

			length							
		L to the zero power		v						
			freq.							
		T to the zero power		area						
			Hz.							
		resis-tance		speed						

(L^3 and L/T^2) and (M and Q) would be in the dimension 3 area, while M/T and Q/T would be in the dimension 4 area.

Dim. -2	Dim. -1	Dim. 0	Dim. 1	Dim. 2	Dim. 3	Dim. 4	Dim. 5	Dim. 6	Dim. 7	Dim. 8
			L		L^3					
		1		L^2		M/T				
			1/T		L/T^2					
		unity		L/T		Q/T				
			length		**M**					
		L to the zero power		v		mass flow				
			freq.		**Q**					
		T to the zero power		area		charge flow				
			Hz.		a					
		resis-tance		speed		flow				
					m					
						cur-rent				
					mass					
						volt-age				
					charge					
						PSI				
					accel-eration					
					vol-ume					

155

Relational Matrix

One could take any term or concept from Physics or engineering, break it down into the resulting M, L, T, and Q units and they would fit on a 11 dimensional matrix from dimension -2 through dimension 8.

Dim -2	Dim -1	Dim. 0	Dim. 1	Dim. 2	Dim. 3
μ_o	T		L	L^2	L^3
		1		area	
perme-ability			1/T		L/T^2
	henry	unity		L/T	a
ML/Q^2			length		
	L induc-tance	L to the zero power		v	M mass
henry/meter			freq.		
	ML^2/Q^2	T to the zero power		speed	Q charge
			Hz.		**E**
	C capaci-tance	resis-tance		**J**	
ε_o		Z imped-ance		Q/L^2T	electric field
	farad		**B**	current density	
$Q^2T^2/$		ρ			$ML/$

156

D-2	D-1	D0	D1	D2	D3
ML^3					QT^2
	Q^2T^2/ML^2	rho	magnetic field		
Permittivity		charge density			volume
			M/QT		
farad/meter		Q/L^3	tesla		acceleration

Dim. 4	Dim. 5	Dim. 6	Dim. 7	Dim. 8
	mv		Fs	
M/T		ma		IxV
	p		W	
Q/T		F		P
	p = mv		W = Fs	
mass flow		F = ma		P=IV
	ML/T		ML^2/T^2	
charge flow		ML/T^2		ML^2/T^3
	momentum		work	

flow		force		power
	Φ_E		torque	
current				H.P.
	electric flux		energy	
voltage		**M** or μ_m		watts
PSI		mag-netic dipole moment		KW
QL				
P				
electric dipole moment		QL^2/T		
D4	D5	D6	D7	D8

If you had two terms in an equation and they were multiplied together, their dimensional aspects would add (like exponents).

F = ma

Force = (mass) (acceleration)

F = m (mass of dimension 3) times a (acceleration of dimension 3)

F (force) would then be in dimension 6.

If you had two terms in an equation and them divided each other then their dimensional aspects would subtract (like exponents).

$I = P/V$

current = power/voltage or

flow = power/pressure

$I = P$ (power of dimension 8) divided by V (voltage of dimension 4)

I (current) would be dimension 4.

The other great thing about this is that it tells you that flow and pressure are of the same dimension. Another way of thinking about it is that they are a manifestation of the same underlying thing.

The end result is a eleven dimensional matrix where all of the terms and concepts of the physical world are shown in their basic relationships with one another.

Dim -2	Dim -1	Dim. 0	Dim. 1	Dim. 2	Dim. 3
μ_o	T		L	L^2	L^3
permeability	time	1	length	area	volume
ML/Q^2		unity	$1/T$		L/T^2
	henry		freq.	L/T	a
henry/meter			Hz.		

	L inductance	L to the zero power		v speed	**M** mass **Q** charge
ε_0 Q^2T^2/ML^3			**B**		
permittivity	ML^2/Q^2	T to the zero power	magnetic field		ML/QT^2
farad/meter			M/QT	Q/L^2T	**E**
	C capacitance	resistance		**J**	electric field
				current density	
	farad				Φ_B
		ρ			magnetic flux
	Q^2T^2/ML^2	rho			
		charge density			
	σ				
	surface charge	Q/L^3			acceleration
	Q/L^2		tesla		
D-1	D-2	D0	D1	D2	D3

160

Dim. 4	Dim. 5	Dim. 6	Dim. 7	Dim. 8
	mv		Fs	
M/T		ma		IV
	p		W	
Q/T		F		P
	p = mv		W = Fs	
mass flow		F = ma		P=IV
	ML/T		ML^2/T^2	
charge flow		ML/T^2		ML^2/T^3
	momen-tum		work	
flow		force		power
	Φ_E		torque	
current				H.P.
	electric flux		energy	
voltage		$\mathbf{M}$ or $\boldsymbol{\mu_m}$		watts
			calorie	
PSI		mag-netic dipole moment	BTU joule	KW
QL				
$\mathbf{P}$				
electric dipole moment		QL^2/T		

ABOUT THE AUTHOR

Jay Hooper has taught for over 25 years in the North Carolina community college system and has worked in industry for 15 years. He has served as a consultant on various projects for over 30 years. He is currently involved with several endeavors with business and industry from his home base in Salisbury, NC through Freeu.

A selection of con-ed and lab manual titles by the author as of July 2014:

Basic Hydraulics: Fluid Power Workhorse

by Jay F. Hooper

2012, 190 pp, paper, ISBN: 978-1-59460-835-3 Electronic Teacher's Materials available

----

Basic Pneumatics: An Introduction to Industrial Compressed Air Systems and Components, Revised Printing

by Jay F. Hooper

2013, 118 pp, paper, ISBN: 978-1-61163-411-2

Introduction to Industrial Motor Control

by Jay F. Hooper

2009, 170 pp, paper, ISBN: 978-1-59460-620-5

Introduction to PLCs, Second Edition

by Jay F. Hooper

2006, 120 pp, paper, ISBN-10: 1-59460-331-6, ISBN: 978-1-59460-331-0

www.ingramcontent.com/pod-product-compliance
Lightning Source LLC
Chambersburg PA
CBHW051509170526
45166CB00001B/451